The Discrete Charm of the Machine

The Discrete Charm of the Machine

Why the World Became Digital

Ken Steiglitz

Princeton University Press / Princeton and Oxford

Published by Princeton University Press
41 William Street, Princeton, New Jersey 08540

In the United Kingdom: Princeton University Press
6 Oxford Street, Woodstock, Oxfordshire OX20 1TR

press.princeton.edu

Library of Congress Control Number: 2018936381

ISBN 978-0-691-17943-8

British Library Cataloging-in-Publication Data is available

Editorial: Vickie Kearn
Production Editorial: Leslie Grundfest
Text Design: Chris Ferrante
Production: Jacquie Poirier
Publicity: Sara Henning-Stout
Copyeditor: Jennifer McClain

This book has been composed in IBM Plex

Printed on acid-free paper. ∞

Typeset by Nova Techset Pvt Ltd, Bangalore, India

Printed in the United States of America

10 9 8 7 6 5 4 3 2 1

To my daughter Bonnie

Contents

To the Reader

What this book is about

The machines we call computers have reshaped our lives, and may in the end transform humanity itself. The revolution is based on just one idea: build devices that store and manipulate information in the form of *discrete bits*. My aim in this book is to explain why this seemingly simple idea is so powerful.

It happens without trying that in pinpointing the virtues of the discrete, digital form, questions arise about the *limits* of the spectacular progress in technology we've seen in the past half century. Computers are cramming more and more components into smaller and smaller spaces, operating faster and faster. Can this go on forever? Computer programs are getting more and more clever. Are there problems that will always be beyond the reach of computers? Will computers become more clever than we? Will they replace us?

At the end of the book, we return to the opening theme and pose a further fundamental question: Will digital computers *always* be superior to analog computers, which use information in continuous, nondiscrete form, or is there some "magic" that remains hidden in the analog world, beyond the reach of the digital computer? The human brain uses both digital and analog forms of information—is Nature keeping some secrets to herself about the ultimate nature of computation?

Who is the intended reader?

Briefly: My ideal reader is interested in science generally, perhaps computers in particular, but is not technically trained. And she just might be curious about why computers are *digital*. This book is not by any means an introduction to computer science, nor is it about how to program or use computers. There are no equations and no computer code. The reader will not escape, however, without some knowledge of how today's computers are built at the most basic, microscopic level, and an appreciation of why they got that way.

A quick tour

There are a number of reasons why computers are digital. Some are physical in nature, and these naturally tend to be more concrete and intuitively clear. For example, the inevitable presence of noise, everywhere in nature, tends to obscure information. Similarly, electrical current consists of the flow of discrete particles called electrons. This means that electrical signals are, at the microscopic level, necessarily granular. We begin, in part I, by discussing these physical obstacles to reliable computation and how they are circumvented by storing and using information in digital form.

We next show how the familiar notion of a valve can provide a building block for all computation. The transistor is a valve in silicon, and the explosive development of solid-state technology reflected in Moore's law has given us the integrated circuit chip that today holds more than a billion transistors. We shall see that the limits of this progress will ultimately be determined by quantum mechanics and, in particular, by Heisenberg's uncertainty principle.

Part II is devoted to two fundamental ideas that emerge from the study of communication rather than physics. Their development resulted in digital signal processing, high-speed networking, and the internet. The resulting ability to share sound and images almost instantly across the globe has changed our lives profoundly in just one generation.

The first idea, Fourier analysis, tells us that we can treat any signal as being composed of a collection of different frequencies. This insight leads to Nyquist's principle, which determines just how fast we need to sample audio and video signals to preserve all their information, and is behind the concept of bandwidth, now a commonly recognized—and critical—resource in our modern world.

The second idea is the use of coding to protect information in a noisy environment. The empirical practice of using redundancy for safely transmitting and storing signals inspired an elegant and influential theory of information, which sprang fully fledged from the brain of Claude Shannon. The crown jewel of the theory is his remarkable (and surprising) noisy coding theorem, which reveals the full depth and significance of the concept of bandwidth.

In part III we move on to yet more sophisticated and challenging territory, ending up, in fact, at the limits of current scientific knowledge. Returning to analog machines for computation, we develop the notion of a problem that is intrinsically difficult. At this point we get a taste of modern complexity theory, the concept of an NP-complete problem, and the most important open problem in computer science.

Finally, we ask if there might be ways to escape the limits of the computers we use today. This leads naturally to the Church-Turing thesis, which asserts that the hypothetical machine invented by Alan Turing essentially captures the concept of computation; and the extended Church-Turing thesis, which takes this one step further, proposing that the Turing machine is the embodiment of all *practical* computation (including analog). We will see that neither thesis is purely mathematical in nature, and neither can ever be proved. From here it is a short step to questions about the ultimate power of computers that take advantage of quantum mechanics.

In the concluding chapter, which constitutes part IV, we review the six main ideas that, in barely a half century, transformed our information technology from analog to digital and led to today's packet-switched and optically delivered internet. We arrive at the edge of the unknown: Are NP-complete problems intrinsically difficult? (Probably yes.) Do Turing machines capture the notion of all *practical* computation? (Probably yes, with a quantum-mechanical upgrade.) Can machines be conscious, and can they suffer? (Quite up in the air.) Whatever the answers to these questions, and regardless of whether their brains will be able to tap unknown analog or quantum power, the current accelerating development of discrete machines is attending the birth of autonomous robots. Ready or not, the robots are coming! How will we face our responsibility to our heirs and successors? Will our human cultural values survive?

A personal note

I grew up on masterpieces of popular science like Gamow (1947), Courant and Robbins (1996), and, later, Russell (2009) and Feynman (2006). These books share one essential feature: they simplify and at times may cut corners, but they never, ever lie. As Ralph Leighton says to the hypothetical student in his preface to Feynman (2006), "There is nothing in this book that has to be 'unlearned.'" With this book I have tried, with my limited resources and all the humility I can muster, to follow these heroes.

Finally, I must confess to a nostalgic attachment to the analog/digital theme. I was born at just about the same time that the first functioning digital computers were being built, but I grew up listening to remarkably practical analog radio. My first paycheck was for summer work writing assembler code for a vacuum-tube digital computer, but I used analog computers in some of my undergraduate courses. My dissertation was on the correspondence between analog and digital signal processing. Throughout this book-in-sonata-form, we develop the analog/digital theme from a variety of points of view; I invite you to chapter 1, the exposition section.

Acknowledgments

There is always the fear that acknowledging the aid or inspiration of colleagues and friends will implicate them in my mischief. Nevertheless, I thank the following for help of all sorts; they are appreciated and blameless: Andrew Appel, Sanjeev Arora, David August, György Buzsáki, Bernard Chazelle, David Dobkin, Mike Fredman, Jack Gelfand, Mike Honig, Andrea LaPaugh, Kai Li, Richard Lipton, Christos Papadimitriou, Mona Singh, Olga Troyanskaya, Kevin Wayne, Andy Yao.

For encouragement, expert guidance, and good cheer, I am also indebted to the staff at Princeton University Press, including acquisitions editor Vickie Kearn, production editor Leslie Grundfest, and editorial assistant Lauren Bucca, as well as to copyeditor Jennifer McClain.

Part I

A Century
of Valves

1 The Discrete Revolution

1.1 My Golden Age of Garbage

What is usually called the "computer revolution" is really about much more—it's about a radical conversion of our view of the world from continuous to discrete. As for your author, my entrance into this world couldn't have been timed better to observe the apparently sudden transformation. I arrived in 1939, a few months before Hitler invaded Poland. At that time the stage had been set, rather subtly and gradually, for the development of things digital, and the pressure of the ensuing war years propelled us all, not so subtly and not so gradually, into what we now know as the Digital Age. This book is about the most basic ideas and principles behind the change. Why did the world change in such a fundamental way from analog to digital, and where might we humans—a species itself built along both analog and digital lines—be headed?

I apologize for the rather dark beginning, but it's a fact that the dirty fingers of war have never failed to leave their prints on the annals of what we term "progress." The dawn of the computer age is closely linked to decryption efforts in World War II, as well as to the development of the atomic bomb.

On August 6, 1945, I was only dimly aware of the fact that I was in New Jersey and not Japan, where bombardier Thomas Ferebee was watching Hiroshima's Aioi Bridge in the crosshairs of his Norden bombsight. The bombsight, which subsequently released the first uranium-fission atomic bomb and began the end of World War II, was an analog computer. It solved the equations of motion that determined the path of the bomb, using things like cams and gears, a gyroscope, and a telescope, all mechanical devices. But it was a computer nevertheless, although applying the term to a mess of moving steel parts might surprise some people today.

Well into the 1950s there were two kinds of computers: analog and digital. In fact, analog computers of the electronic sort were the only way to solve certain kinds of complicated problems, and were, in a handful of situations, very useful. Electronic analog computers were programmed by plugging wires into a patch panel, which was like a telephone switchboard (you may have seen one in an old movie), and by the time any interesting problem was running, the patch panel was a rat's nest.

But before the mid-twentieth century *everything* was analog; digital just hadn't been invented.[1] The most important piece of information technology I knew as a child was the radio, very analog at the time, and it was my remarkable piece of good fortune when the postwar engines of production turned to consumer goods, and consumers bought new, streamlined, plastic radios. Garbage night meant that the monstrous mahogany console radios of the 1930s could often be found curbside—with booming bass, hardly any treble because of the limitations of AM broadcasting, and all manner of interesting electronic parts inside.[2] That was how I learned to love the glow of vacuum tubes and the aroma of hot rosin-core solder congealing around the twisted leads of condensers (as capacitors were called), resistors, coils, and other more exotic components. Sometimes it was an autopsy that I performed on these found radios, but often it was a vivisection, since many of them worked, or could be made to work, excellently. Some of these lucky finds even had shortwave bands, and garbage night turned out to be my gateway to the world at large.

It was all analog. When television came, that, too, was all analog. So were telephones. There just wasn't anything else.

1.2 Nostalgia and the Aesthetics of Technology

Video and audio signals fly in and out of our brains all day long, and devices that process those signals—radio, television, recorded film and music players, telephones—were all digitized in the latter half of the twentieth century; that is, within my lifetime. One consequence is that the devices we use every day for what is now called digital signal processing have more or less converged to the same, rather dull-looking machine—essentially a small chip behind a screen, in a plastic case, occasionally with a couple of wires hanging out. In contrast, in the good old days radios were *radios*, television sets were *television sets*, cameras *cameras*, telephones *telephones*. You could tell what a device did by looking at it. And sometimes you would need an elephant to make it portable: the Stromberg Carlson console radio I lugged home with the help of my friends was crafted with a sturdy wooden cabinet, housing a loudspeaker with a huge electromagnet, a large lit dial, and hefty knobs that gave the operator the feeling of controlling an important piece of equipment—to a child, and perhaps to a grown-up as well, a spaceship.

My favorite effect was the *magic eye* tuning indicator, usually a 6E5 vacuum tube that had a fluorescent screen at its end, visible in a circular hole on the front panel of the radio. It glowed green with a dark crescent that contracted in proportion to the signal strength. Carefully tuning a station to reduce the crescent to a narrow slit was a joyful experience, especially in a dark room where the eerie glow did seem magical for sure. Punching in the frequency (or URL) of a radio station just does not provide the same tactile and visual pleasure. If your childhood came after such electronic apparatus, you don't know what I'm talking about; such is the nature of nostalgia. No doubt the iPhone will stimulate similar feelings fifty years from now, when signals may very well go directly to our brains without the need for any beautiful little intermediary machines.

Of course there is a lively market for retro style and retro devices; certain cults have grown around the disappearance of, for example, shellac, vinyl, and analog tape recordings, or film cameras and the once pervasive technology of chemical-based photography. It's common to hear that vacuum-tube amplifiers have a "warmer" sound, although it's not certain how much of the warmth is due to distortion from the inherent nonlinearity of the vacuum-tube analog technology, or the psychological glow from the hot tubes themselves.

Sometimes the nostalgic longing approaches the mystical. Water Lily Acoustics produces superb recordings of Indian classical music, and they go through great pains to keep the sound recording free of the digital taint until the very last step in the process. For example, the booklet for a compact disc recording of Ustad Imrat Khan offers the following assurance:[3]

> This is a pure analog recording done exclusively with custom-built vacuum-tube electronics. The microphone set-up was the classic Blumlein arrangement. No noise reduction, equalization, compression, or limiting of any sort was used in the making of this recording.

The booklet goes on to describe the microphones (which use tubes), recorder (Ampex MR70, half-inch, two-track, 15-inch-per-second tape, using vacuum tubes called *nuvistors*), and so on.

Spiritual values aside, a good analog sound recording, or, for that matter, a good analog photograph taken with film and printed well, can be, technically, a lot better than a bad digital recording or a bad digital photograph. We have much more to say about the ultimate and practical limitations of analog and digital technology as we go along.

1.3 Some Terminology

So far, we've been using the terms *digital* and *analog* rather loosely. Before going further, we need to clarify this terminology. For our purposes, *digital* means that a signal of interest is being represented by a sequence or array of numbers; *analog* means that a signal is represented by the value of some continuously variable quantity. This variable can be the voltage or current in an electrical circuit, say, or the brightness of a scene at some point, or temperature, pressure, velocity, and so on, as long as its value is continuously variable. All the possible values of a digital signal can be *counted*, and there is a definite gap between them; those of an analog variable cannot be counted, and there is no definite gap between them. Generally, we use *discrete* (actually "discrete-valued") to mean digital and *continuous* (actually "continuous-valued") to mean analog, although this overlooks some distinctions that are not important at this point.

When you buy a wristwatch or a clock, for example, you have a choice between an "analog display" and a "digital display." This is exactly the sense in which we use the terms—but take note of the fact that we refer to the *display* and not the internal mechanism of the timekeeper. A clock with an analog display has hands that can move continuously, whereas a digital display shows numbers that change discontinuously, which is another way to say suddenly. The hands of a clock actually represent time by the rotational position of gears. These days, the usual clock with an analog display has an internal timekeeping mechanism that is digital (except for old-fashioned windup clocks). But at one point there were the opposite kinds of clocks, with analog mechanisms and digital displays—usually using gears and cams to flip displays with numbers printed on them.

On the morning of "Pi Day" (March 14) of 2015, there was a moment a bit after 9:26 and 53 seconds when the time could be written 3.14159265358979...; that is, π. The moment was fleeting to say the least; it was infinitesimally brief. And it will never occur again. Ever. If you were watching the hands of a clock with an analog display, you might have tried to take a photo at the exact moment of π, but the photo would have taken some finite time, and you would have necessarily blurred the second hand. That is an inevitable consequence of measuring an analog quantity of any kind.

Very commonly, audio and video signals are represented by voltages, either in a computer, smartphone, copper cable, or some kind of electrical circuit like those in an amplifier. This is the usual way that such signals are recorded by microphones and video cameras, and the resulting signals are transmitted and reproduced using voltages in electrical circuits. A microphone converts a sound pressure wave in the air to a time-varying voltage. A video camera converts a light image into an array of time-varying voltages. These audio and video signals usually start their lives out as analog signals and are converted to digital form after their initial capture, assuming that they are going to be processed in some way in digital form.

The device that converts an analog signal to digital form is called, naturally, an *analog-to-digital converter* (*A-to-D converter*), and the opposite operation is performed by a *digital-to-analog converter* (*D-to-A converter*). Thus, for example, the light-sensitive screen in a digital camera is really an A-to-D converter, whereas your computer monitor is really a D-to-A converter.

I'll try to be clear about what I mean when we use the terms *digital, analog, discrete,* and *continuous,* but I should mention some possible sources of confusion. First, it often happens that it is *time* itself that is thought of as discrete or continuous, rather than the values of a signal. When there is any possible confusion, I will state explicitly that time is being considered. Second, there is the awkward fact that standard mathematical terminology uses the term *continuous* in a slightly different way. Mathematically speaking, a curve is "continuous" if it does not jump suddenly from one value to another but rather changes "smoothly." The reader who has studied calculus will be aware of this alternate interpretation, but will not be confused by it.

Finally, the term *discrete* is used by physicists in another sense. A most important example of this usage comes up when we ask the question, "What is light?" The question has puzzled scientists for centuries. Sometimes light behaves like waves; this is evident when we observe diffraction rings, for example. If we aim a narrow light beam (say, from a laser) through a pinhole, and project the result on a screen, we get concentric rings that die out in intensity as we travel from the center. It turns out that this result is easy to explain if we treat light as a wave but very difficult to explain if we treat light as particles. On the other hand, if we aim a light at a detector and gradually decrease its intensity, eventually the light does not become dimmer and dimmer without limit. At some point the light begins to arrive in chunks: Click! ... Click! You can hear such clicks if you receive the light with a sensitive detection device connected to an amplifier and speaker. This experiment and many others provide evidence that light consists of particles; a wave would fade out, diminishing in intensity indefinitely. The particle of light, called a *photon,* is indivisible. There is no such thing as half a photon or half a click. A click occurs or it doesn't. All the clicks are the same. In such cases we say that light is *discrete;* it occurs as *discrete* particles.

All chunks of matter—atoms, molecules, electrons, protons, and so on—also behave in this same seemingly paradoxical way. The puzzle, sometimes known as *wave-particle duality,* was ultimately explained after a great deal of hard work by some very smart people about a hundred years ago. The explanation is called *quantum mechanics*, which not only revolutionized physics but changed the way we think about the world.

Quantum mechanics, and physics in general, plays an important part in our story, and we return to it often. It is the science of the very small. As put by Jean-Louis Basdevant, "Bill Gates, the richest man in the world, made his fortune because he was able to use [micro- and nanotechnologies]; quantum mechanics accounts for at least 30% of each of his dollars."[4]

More about quantum mechanics later. We next turn to the fundamental role of physical noise in limiting the performance of analog devices, and the way in which digital devices circumvent the problem.

2 What's Wrong with Analog?

2.1 Signals and Noise

We use the words *signal* and *noise* in everyday speech. Opinionated music lovers mean one thing when they speak of noise, insomniacs something different, stock traders yet something else.

Scientists and engineers use the words in the following specialized way: A *signal* is the part of what we perceive that carries information to us. *Noise* is what we perceive that carries no information, but rather tends to obscure the signal. Over the past few decades this slightly more technical usage has diffused into general usage, most notably in economics reportage. For example, reporters on the Federal Reserve Bank's policy pronouncements speak of its signal-to-noise ratio.

Noise is unavoidable in our world, and almost always undesirable, so it is very interesting to think about the differences in how it affects analog and digital signals. Consider, for example, the (analog) voltage in an audio amplifier that represents the (analog) sound pressure wave at the microphone capturing a concert. At some particular time the value of that analog signal might be, say, 1.05674... volts. We can write down that value only to a certain number of decimal places, but in theory the digits can go on indefinitely. Mathematically speaking, numbers like this, representing analog quantities, are called *real numbers*. Noise in the amplifier changes this value as the signal propagates through the circuit. At some point it may be changed to 1.05656... volts, say, due to the addition of a noise voltage of −0.00018... volts. I am being careful to indicate that these numbers are real and cannot usually be written neatly with a finite number of digits. The important point is that an analog signal gets blurrier as it gets corrupted by noise, and in general this is an irreversible process.

Contrast this with the situation of a particular piece (bit) of a digital signal, which can, at any particular point in a computer, take on only the values 0 or 1. If the noise at some point is really gigantic, a 0 might get changed to a 1, or a 1 to a 0. But otherwise its value will be, insofar as we are allowed to assign values to the bit of the digital signal, *exactly* 0 or 1. We have more to say about this crucial difference in the next chapter. It is enough to note here that there is a certain *threshold* below which noise will have no effect at all. It is rare to have an opportunity to use the word "perfect" with complete accuracy, but if we ensure that the noise in a digital machine is below that threshold, the operation of our device will be literally perfect.

2.2 Reproduction and Storage

We have now come to the first important problem with analog signals and the world of analog gadgets. Every time an analog signal is stored, retrieved, transmitted, amplified, or processed in any way, it is unavoidably corrupted by noise. We can be very careful about reducing the amount of noise, but it cannot be reduced to zero. Furthermore, its effects are irreversible and accumulate as we continue to process a signal.

This phenomenon is very well known to those who edit sound. In the old (predigital) days, if you laid down tracks for the performance of a song, merged these tracks, added more tracks, then filtered the results, and so on, eventually you ended up with a product that was just too low quality to use. Every stage of processing added its own noise,[1] and by the time you were done with 10 or 20 stages using, for example, an analog tape machine, you were, sonically speaking, down in the mud. Digital editing does not suffer from this limitation, assuming we use enough of those 0 or 1 bits for each value of the sound signal.

2.3 The Origins of Noise

I've claimed that noise is unavoidable, but I haven't explained why that's so or, in fact, where noise comes from in the first place. There are actually many ways that noise can arise in physical systems, and questions about its nature and inevitability run deep. We're going to develop some answers gradually as we go along.

The simplest place to start is with the fact that the world consists of molecules, atoms, and particles in a state of constant agitation. It's not obvious, but if you could use a super microscope to examine the air above a calm meadow on a warm summer day, or the water in a tranquil pool, or a rock in the pool for that matter, you would see molecules constantly bouncing around. The higher the temperature, the faster the vibration of the constituent particles. When this agitation corrupts the signals we are interested in, it is called *thermal noise*,[2] which is in a sense the most basic and easiest kind of noise to understand.

Thermal noise provided the first direct evidence that matter was discrete, and the process of firmly establishing the atomic theory of matter took about a hundred years. In 1827 the Scottish botanist Robert Brown observed random motion of tiny particles trapped in pollen that was suspended in water. This had been noticed before by others, but he studied the phenomenon very carefully and showed that the motion was not due to any life force in the pollen, as he thought at first. Over the following decades this "Brownian motion," as it is now called, was shown to be caused by the constant battering of the tiny particles by the yet much smaller water molecules. Albert Einstein proposed a mathematical theory that explained Brownian motion in his miracle year of 1905, and Jean Baptiste Perrin verified Einstein's theory experimentally, for which he was awarded the Nobel Prize for Physics in 1926.[3]

2.4 Thermal Noise in Electronics

Very often we deal with signals and noise in electronic equipment, where they are represented by voltages and currents. Electronic components like resistors very naturally contribute thermal noise, and the power that each component contributes is proportional to its temperature. In this case it is the charge carriers, like electrons, that are subject to thermal agitation.

To take a typical example of an analog electronic device, an analog radio receiver consists of a sequence of stages that amplify the signal that arrives at the antenna, which is measured in microvolts,[4] to the point where it can drive a loudspeaker, where it is measured in volts. Thus, the amplifiers in a radio receiver can easily increase the size of the signal by a factor of a million. Noise that corrupts the signal received by the antenna—the radio-frequency, or RF, signal—is therefore magnified the most, and it is for this reason that controlling the noise in the earliest stages of an analog device (the front end) is most important. If you're old enough to have seen "snow" in the picture of an analog television set receiving a broadcast signal, it was the manifestation of noise added to the signal itself on the way to the antenna and in the earliest stages of amplification.

Astronomers worry a lot about minimizing noise in their imaging systems, whether they are collecting radio or optical signals. It is therefore quite common for them to cool their electronic detection equipment to extremely low temperatures using liquid nitrogen or helium.

2.5 Other Noise in Electronics

Thermal noise isn't the only kind of noise in electronic equipment. *Shot noise* occurs because the flow of electricity is carried by discrete particles, typically electrons, and electrical current is therefore corpuscular, like the arrival of photons in a light beam, or the dropping of sand in an hourglass. Most of the time, of course, the granularity isn't noticeable because the charge of each electron is so small. A current of one ampere, used by a bright incandescent lightbulb, for example, corresponds to roughly 6×10^{18} electrons per second—about a billion electrons for every human on earth, every second.

The granularity of shot noise is determined by the fixed charge of the electron, so the smaller the current the larger the *relative* size of shot noise. Currents in integrated-circuit transistors are vastly smaller than the one ampere in the lightbulb mentioned above, and might be measured in microamperes (10^{-6}), nanoamperes (10^{-9}), or even picoamperes (10^{-12}). Such currents will then be far from the torrents of electrons in our power supply, and more like "rain on a tin roof."[5] Horowitz and Hill (1980) illustrate the point with an example where the current is one picoampere, in which case the relative size of the shot noise is more than 5%, hardly negligible and potentially quite damaging.[6]

Another kind of noise that appears in electronic equipment is *burst noise*, which appears as sudden and randomly occurring jumps in voltage or current. It is also called "popcorn" noise, because on a speaker it sounds like cooking popcorn. The causes of this kind of noise, and there are more than one, have been attributed to defects in semiconductor devices, especially defects in semiconductor crystals. Burst noise is minimized by quality control in the manufacturing process and by testing after manufacture, when especially noisy devices can be culled. In a sense this source of noise is less fundamental than thermal or shot noise.

Finally, we mention *1/f noise*, or *flicker noise*, or *pink noise*, which is more difficult to explain than any yet mentioned, but nevertheless important and potentially troublesome. For this we need to introduce the idea of a *spectrum* or *power spectrum*. We can think of signals or noise as being composed of constituent frequencies. An old-fashioned radio dial, for example, shows the available frequencies of electromagnetic waves that the radio can receive. A prism splits white light, which contains all frequencies in equal amounts, into a rainbow of its constituent frequencies (colors), spread out linearly like a radio dial. Red is at the low-frequency end of the visible spectrum and violet at the high-frequency end.

Thermal, or Johnson, noise is "white." Its spectrum is flat, meaning that all frequencies are equally represented. Actually, the idea that any signal or noise can be thought of as a combination of different frequencies—Fourier analysis—is very general and fundamental, and is an essential tool in many areas of science and technology. Fourier analysis is named after Jean-Baptiste Joseph Fourier, who used it to solve the important problem of how heat diffuses. The idea that certain waveforms can be broken down into constituent frequencies was in the air well before this work, but Fourier took the crucial, if not quite rigorously justified, leap of assuming that *any* waveform can be resolved into its component frequencies. We meet Fourier analysis again when we discuss signal processing in chapter 6.

Now, when Johnson measured thermal noise, which was in the mid-1920s, he noticed extra power at low frequencies. Today this additional contribution is sometimes called *excess noise* because it adds to the thermal noise. He found that the power of this excess noise at low frequencies was inversely proportional to the frequency, hence the name *1/f noise*, f being the frequency. This turns out to imply that the power in every frequency decade is the same: the total power between 100 Hz and 1000 Hz is the same as the power between 10 Hz and 100 Hz, and between 1 Hz and 10 Hz, and so on. Light with this kind of distribution of frequencies appears pink, hence the alternate name *pink noise*. One way to look at the fact that every decade has equal power is to say that randomness occurs on all timescales—there are noise components that change at all rates, from very slowly to very quickly.

Our consideration of $1/f$ noise brings us to an interesting conundrum: If you add up all the power at lower and lower frequencies, you get infinity! This is sometimes called an *infrared catastrophe*. You can see why this is true quite easily from the fact that there is equal power in every frequency decade. Suppose the decade from 100 Hz down to 10 Hz contains a certain amount of power, say, P. (We can start at any frequency for this argument.) Now add to that the power from 10 Hz down to 1 Hz, which gets us to $2P$. Then from 1 Hz to 0.1 Hz, getting to $3P$. If the power spectrum really is $1/f$ all the way down to 0, this process can go on forever, so the total power grows without bound. Very disconcerting . . . how can a little resistor (say) with its seemingly innocent $1/f$ noise, generate an infinite amount of noise power?

Notice that the same problem occurs if we add up the total power going *up* in frequency, an *ultraviolet catastrophe*. But ridiculously high frequencies in real systems can be argued away easily on physical grounds (for one thing, we would all be killed by the radiation), whereas the presence of very low frequencies is more difficult to dismiss.

The puzzle attracted wide attention as power spectra that behaved like $1/f$ at low frequencies started showing up in a wide variety of fields besides electronics. The excellent review papers of Milotti (2001) and Press (1978), for example, describe closely related spectra measured in ocean current velocity and sea level at Bermuda, earthquakes, sunspot number, the light curves of quasars,[7] and loudness and pitch fluctuations in both voice and music broadcasts, including Scott Joplin piano rags, classical music, rock, and news-and-talk.[8]

When physicists and mathematicians see the same kind of behavior popping up in seemingly unconnected areas, they get smiles on their faces and they go to work looking for an underlying explanation. As Milotti (2001) puts it, "The appearance of power laws . . . seemed to indicate that something deeper was hidden in those ubiquitous spectra." Press (1978) shows a plot of the spectra of $1/f$ noise in three different electronic devices— a carbon resistor, a Germanium diode, and a vacuum tube—and he remarks, "The frightening thing about [the figure] is that the noise spectra shown are still rising along a beautiful power law at the smallest frequencies measured." I don't know how frightened we should be, but the increasing power at very low frequencies means that the noise is correlated over very long periods of time. And yet careful measurements over longer and longer times (corresponding to lower and lower frequencies) in many different areas have not revealed a leveling off of the spectrum as the frequency approaches zero. The puzzle remains: How does a resistor (for example) remember its place in a voltage fluctuation that may take weeks or months?

Neither Milotti (2001) nor Press (1978) come to any conclusions about whether $1/f$ noise has a deep and general explanation. Milotti sums it up neatly, if a bit anticlimactically, with, "My impression is that there is no real mystery behind $1/f$ noise, that there is no real universality and that in most cases the observed $1/f$ noises have been explained by beautiful and mostly ad hoc models."

The purpose of this little excursion to the land of noise is to convince you—or begin to convince you—that noise in physical systems is inescapable. We pile on the evidence later, but now we're ready to see how noise makes it difficult to deal with infor-mation in analog form.

2.6 Digital Immunity

It's catchy technical jargon to speak of "software rot," but of course software can't rot; it's text, and it can't rot any more than a Shakespeare play can rot. The way we use it, change it inadvertently, or change it without taking full account of the consequences can cause software to malfunction or degrade in performance. But if you load the same software into the same machine with the same initial state, it will always do the same thing: it runs *deterministically*. Later, we'll see that the world itself does not behave deterministically, because of noise and the even more fundamental phenomena of quantum mechanics, and it is quite remarkable that we can do things that are, for practical purposes, deterministic in a nondeterministic world.

Software and text are examples of information in digital form, and they can be stored, transmitted, and retrieved perfectly. Some may take issue with such an unqualified statement. There is, after all, some very small chance that all the molecules in, say, a solid-state memory will suddenly wander into the next room, but the probability of this happening in a lifetime has a number of zeros after the decimal point that is staggering. Furthermore, the chances of corrupting the data can be made as small as we wish by introducing redundancy, or some kind of fancier coding (at some expense, of course), something we discuss further in chapter 7. For this reason we should claim, more properly, that the probability of losing any digital information can be made vanishingly small.

Eventually, you say, any medium will deteriorate, and stored data will therefore always be lost in the long run. But because digital data can be copied perfectly, it can be transferred to a fresh medium, often newly invented. I'm reminded of a program for A-to-D and D-to-A conversion that was written at Princeton University in the early 1970s for what was then called a "mini-computer" (specifically, a Hewlett-Packard 2100A). The standard program input and output medium for this machine was punched paper tape; the floppy disk was not yet available. Now this machine was used almost exclusively for D-to-A conversion of music, using gigantic digital tape drives, but the conversion program was always read in from the paper-tape reader.[9] Paper tape is made out of . . . paper. Wear and tear occurred quite literally and many backup copies were always on hand. Mylar tape improved life considerably. It seemed practically indestructible (I couldn't tear it with my hands) and copies of the program still survive. Of course if you wanted to use such a program today, you would want to find a paper-tape reader to transfer the contents of such a tape to more modern media, but, push come to shove, and in desperation, the holes are 1.83 mm in diameter and can be read easily by eye.[10]

You can imagine a sequence of transfers over decades from one medium to the next, progressing from a punched paper tape, to a punched mylar tape, to a floppy disk, a compact disc, a flash drive, DNA, and to who knows what next—but the code of the stored program would remain exactly the same.

Consider television, another example of the benefits of essentially incorruptible digital signals. These days we're spoiled by the digital images delivered to our homes by cable or satellite signal. The bits either arrive correctly or they do not, and when it works, it works very well indeed. Failures are usually catastrophic, and usually necessitate a call to our service provider. Analog television is vanishing, but older readers will remember the never-ending struggle with ghosts (fainter duplicates of images displaced from the original) and snow, the video rendering of thermal noise that we discussed above. Similarly, digital radio and music now provide wonderfully clean sound, or nothing at all. It wasn't too many years ago that the crackle and pop of AM radio static and LP records, and the hiss of fringe-area FM reception were unavoidable parts of listening life.

Incidentally, the broadcasting of a video or audio signal can be thought of as a form of storage: the signal is stored in electromagnetic waves and retrieved by the receiver. This may seem a far-fetched interpretation here on Earth, where radio signals can get anywhere in a small fraction of a second; but on average it takes a radio signal, which travels at the speed of light, about 79 minutes to travel from the Earth to Saturn. This means that a broadcast of Beethoven's Ninth Symphony could be stretched out in space between here and Saturn, stored in the form of electromagnetic waves in space, instead of, say, an (analog) vinyl disc or a (digital) compact disc. This fantasy assumes we have enough power and big enough antennas for reliable transmission and reception at the CD rate, something we'll leave to NASA to worry about. By the way, light travels 186,000 miles per second, and the data rate from a spinning compact disc is 176,400 bytes per second, so the symphony stretching between here and Saturn would take up just about one byte every mile.[11]

Of course the ultimate example of the purity of digital signals is around us everywhere in our modern world, in all the devices that are essentially computers—whether we call them that or not. Your desktop, laptop, smartphone, digital watch, GPS tracker, automobile engine control unit, and digital camera are all processing digital information, performing millions of logical operations per second, with essentially no (hardware) errors at all.

2.7 Analog Rot

Analog processing is a different story. The information in an analog signal is represented by a physical quantity, and physical quantities are always subject to some kind of deterioration, whether they be voltages, currents, silver halide crystals in gelatin (film), electromagnetic waves, grooves in shellac or vinyl, magnetized particles of an iron oxide on a tape, or any of the many media used today or in the past. In the same way, the machines we use to convert analog information from one form to another are always subject to the imperfections of the real world. Phonograph turntables add rumble to the sound because there is always some coupling from the motor to the table, wow because the speed is never perfect, distortion because the stylus never follows the groove perfectly, hum because there is always some coupling between 60-cycle power lines and the low-level audio cables, and so on.

The discussion of electronic noise in sections 2.4 and 2.5 illustrates some of the most obvious and common manifestations of analog noise, but we should not lose sight of the fact that all analog signals are fundamentally mortal. As a case in point, many early motion picture films, especially silent films, have been completely lost because of deterioration of the film stock. In fact, nitrate film stock, used until the 1950s, is highly flammable and if not stored and handled properly can be downright dangerous.

Once an analog signal of any kind is corrupted by noise, the damage is generally irreversible. It may be possible to ameliorate the effects of scratches on a phonograph record with filtering of some kind, or patch a torn motion picture film, but cumulative wear and the effects of aging are always permanent. The problem of restoring old sound recordings and motion picture films is a vast and sometimes controversial subject, and we have more to say about it later, when we discuss digital signal processing.

Before we return to our main program . . .

2.8 Caveats

I hope it's clear that we are dealing with general principles and ultimate trends, not the details of any particular application at any particular time. I can't argue with the audiophile whose state-of-the-art system gets the best sound ever heard from a vinyl recording (never played before, because phonograph styli wear records out), and an analog amplifier with a pair of golden output tubes, driving a 40-pound transformer. Or the photographer who captures a scene on film that, for one arcane reason or another, would have been impossible to get with a digital camera. There may even be some applications where an analog computer may, at any given time, outperform any digital computer. I also can't deny the aesthetic appeal of a highly evolved, antique machine. I've already acknowledged it and I readily succumb to it. But technology moves on, and my primary goal in this book is to explain why digital information processing is triumphing over the analog alternative, and is likely to continue to do so. Be aware, however, that in part III we examine the highly speculative—but entertaining—notion that, at least for certain critical problems, final victory might just go to the analog machine.

I should also qualify my claim that data stored in digital form is immortal. This is true only if a program for refreshing data (perhaps to new media) is in place, since all data is ultimately stored in physical media and therefore vulnerable to decay. Furthermore, any transfers involved in such refreshment must be what is called *bit faithful*, preserving all the original information, and that is not always the case. For example, compact discs are encoded with redundant bits to permit their being played even if the surface of a disc is scratched. If we copy copies of copies imperfectly, indefinitely, we will ultimately lose everything. But the availability of relatively easy encoding is one of the strengths of the digital idea, and makes modern digital communication possible, as we shall see in chapter 7.

3 Signal Standardization

3.1 A Reminiscence

Professor Arthur Lo's office was down the hall from mine at Princeton in the 1960s, and from time to time I would wander by to see what he was up to. He had a reputation for being a man of depth, a thinker, and he was always happy to chat with a young newcomer. I would usually find him chomping on his pipe, peering out from a bluish haze of smoke. His important talent was boiling apparently complicated matters down to a simple and elegant form. He told me about two principles that make digital computers possible: *signal standardization* and *directivity of control*, which today might be considered "obvious," but these were early, formative times for computers. Nothing was obvious. I've been happy to carry these ideas with me ever since.[1]

3.2 Ones and Zeros

The *defining characteristic* of the digital medium is simply that a signal bearing information can take on only a discrete number of values. In this case the simpler the better, and it turns out that it's possible to get by very well indeed with only two distinguishable values, *bits*, which are conventionally called "1" and "0," or "TRUE" and "FALSE," or "ON" and "OFF," depending on the context. These are often referred to as *logical values*, to distinguish them from the values of analog, physical quantities used to represent them. In a real electronic circuit, TRUE and FALSE might be represented at any particular point by 5 and 0 volts, or +2 and −2 volts. It doesn't matter, as long as we use two different, reliably distinguishable values. This is why computer scientists count with two fingers instead of ten, using binary, or base 2, arithmetic instead of base 10. Digital technology, and the reason it exercises such dominance over analog alternatives, stems from having discrete alternatives for signal values—base 3 or base 17 would also work, but not as elegantly.

In some situations, however, there is a real advantage to using only two possible values. If we do, we can use "POSITIVE" and "NEGATIVE" values of some physical signal to represent bits; or even the "PRESENCE" and "ABSENCE" of some signal.

The crux of the matter occurs where analog and digital meet. The "real world" is analog,[2] so how can we ensure that signals can take on only two possible values? The answer to this question is found in the natural process called *signal standardization*.[3] Let's say bits are represented in some typical electronic circuit by the nominal values 5 volts for TRUE and 0 volts for FALSE. These are analog values, and, as we know, they get corrupted by electronic noise. At some time a TRUE signal may actually be 5.037 volts. At another time the same TRUE signal may be 4.907 volts. A FALSE signal may at some time be 0.026 volts and at another time 0.054 volts. The circuits that handle these values are called *digital logic circuits*, and they have the absolutely critical property that they push values near 0 volts toward 0 volts and values near 5 volts to 5 volts. As we progress through the circuit, this has the effect of *standardizing* the signals, in the sense that a signal meant to have the logical value TRUE will always be represented by a voltage close enough to 5 volts to be distinguishable from a signal meant to be 0 volts, and vice versa.

The way standardization is accomplished in any particular computer varies with the kind of computer. Most computers today are electronic, and the circuits that standardize signals simply ensure that voltages above the halfway point of 2.5 volts are pushed toward 5 volts and those below 2.5 volt toward 0 volts (keeping our example where 5 volts is "TRUE" and 0 volts is "FALSE"). For an error to occur, there must be some point in a circuit where the noise is larger than 2.5 volts, and in the usual electronic circuits, where the average noise excursion might be measured in millionths of a volt, the chances of this happening are infinitesimally small.

Digital computers can be made out of mechanical parts like gears and cams, or tubes that carry fluids like air or water, for

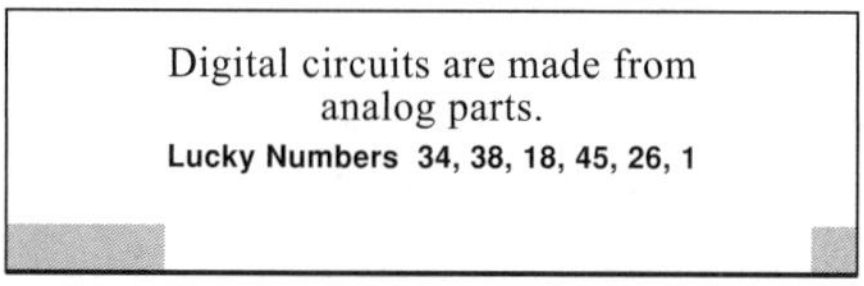

Digital circuits are made from
analog parts.
Lucky Numbers 34, 38, 18, 45, 26, 1

FIGURE 3.1. I really did get this fortune in a Chinese restaurant fortune cookie.

example, but the same principle applies: logical values must be standardized from stage to stage to stage, or maybe every few stages, so that for practical purposes we can always think of the signals that represent them as having discrete values.

The essence of this section is captured by the message shown in figure 3.1. I leave its metaphysical interpretation, if any, to the reader.

3.3 Directivity of Control

I've referred above, without explanation, to logical signals propagating in one direction, from stage to stage, in a digital computer. This brings us to the second critical property that signal-carrying elements must have for a computer to work: signals must be *unidirectional. Controlling elements* must control *controlled elements*, and never the reverse.

According to this picture, we can envision a digital computer as a network of interconnected elements, called *gates*, each of which has controlling logical signals (each TRUE or FALSE) called *inputs*, that determine controlled logical signals called *outputs*. This is not to say that control cannot sometimes loop back on itself. It may very well happen that gate A controls gate B, which controls gate C, which in turn controls gate A. But each gate determines its output from its inputs, and the outputs of each gate can control only the inputs of others.

3.4 Gates

Commonly used gates do commonsense things and are easy to understand. For example, an AND gate has two inputs and one output; the output is ON when and only when both inputs are ON.

We don't need to go into any detail here about the different kinds of gates and how they are used to build all kinds of interesting things like internet browsers and speech synthesizers. Rather, we concentrate on the few very simple structural principles that make digital computers possible—like signal standardization, directivity of control, and rigid control over *when* signals are allowed to change (clocking).

We do have to worry, however, about the potential problem of having to build thousands of kinds of complicated gates, one for each particular application. We avoid this using the principle of *modularity*, something close to the heart of every computer scientist. Without this principle it would be practically impossible to assemble the complex digital computers that we use today. Computers are organized as a hierarchy of modular construction, layer upon layer, building with a very few simple abstractions. And at the bottom level it turns out we can use the *valve*, something we use every time we turn on a water faucet, an act we don't usually think of in abstract terms. If you are delighted by the idea that all you need to build the smart part of your smartphone is a box of identical parts (a billion or two of them!), then you are, at heart, a computer scientist.[4]

That you can build a whole computer from one kind of part is a sweeping, perhaps surprising claim. You might be wondering, for example, about the clocks we need to step the logic from stage to stage, and the memory we need to hold values for future use. But these, too, can be built from valves, using some feedback tricks. The complete story would entail an unnecessary detour. But the story of the valve as a basic digital computer component is worth telling here in a little detail, because it is intimately related to the discovery of the electron and, paradoxically perhaps, the development of the strictly analog technology that shaped the first half of the twentieth century: radio, television, telephone, radar, and all things electronic.

3.5 The Electron

In 1897, just a bit more than one hundred years ago, Sir J. J. Thomson published an important paper showing that the flow of electricity in a vacuum, called *cathode rays*, is actually composed of streams of tiny particles ("corpuscles"), and suggesting that these corpuscles are a basic constituent of all matter.[5] Recall that we have already mentioned this in connection with shot noise. This was a time when the discrete nature of almost everything was being discovered. The paper is generally regarded as announcing the discovery of the electron, and it led to the cracking open of the atom and the unraveling of its structure over the next few decades. It is a pleasure to read even today, a century after it was written. Thomson describes with great clarity his struggles with experimental uncertainties, and his brilliance shines through the pages.

3.6 Edison's Lightbulb Problems

The cathode rays that Thomson was studying come about as follows. If a filament made out of a material that can withstand high temperatures, say, tungsten, is heated in an evacuated glass envelope, electrons boil off its surface. That is, the thermal energy of the electrons allows them to escape the forces that would ordinarily hold them in the filament. When Thomas Alva Edison was perfecting his lightbulb, he used a carbon filament and was plagued by two problems: carbon deposits on the inside of the bulb, evidently caused by carbon leaving the filament; and the attendant problem of the filament thinning and breaking. To alleviate the first problem, he tried introducing a second element in the tube to intercept the carbon deposits. In the process of his endless experimentation, Edison discovered that a current flowed from the filament to this second element. He patented the device in 1883, and the effect is called the "Edison effect." There was other related work on the flow of currents carried in specially fabricated evacuated tubes, and the invention of these tubes is generally credited to Sir William Crookes in about 1875. The nature of the current, however, was quite unclear until Thomson's brilliant 1897 paper.

John Ambrose Fleming applied the Edison effect in 1904, when he surrounded the filament by a cylindrical plate, the positively charged *anode*, and used the device to create a new kind of *diode*. The defining property of such a diode is that electrons can flow through it in only one direction, in this case from the filament to the anode, but not in the reverse direction. This was, in a sense, the first "vacuum tube" and was called a diode because it has two elements within the glass envelope. These days the word *diode* suggests a solid-state device, but for half a century most diodes were made this way, using an evacuated glass tube with a heated filament.

The reason Fleming's device acts as a diode, or one-way valve, is easy to see—a lot easier to see, in fact, than the reason solid-state diodes work. Free electrons can carry charge when they travel from the hot filament to the anode, but there is nothing that can carry charge in the reverse direction. As we've described, it is easy to boil electrons off hot metal in a vacuum, but protons are another matter entirely![6]

3.7 De Forest's Audion

It was only two years later, in 1906, that Lee De Forest took the next step. He inserted a third element in the evacuated tube, a zigzag grating of wire called the *grid*, between the filament and the anode, which could then be used to control the flow of current between those two elements. He called this new three-element device the *audion*, but the generic name has become *triode*. You can see the arrangement of the three elements in the original patent, figure 3.2. Later on, grids would become meshes or coils surrounding a *cathode*, which was heated by a filament, and the general class of devices—glass envelopes in which metal elements controlled the flow of electrons in a vacuum—became known as *vacuum tubes* or, in the United Kingdom, *thermionic valves*.

De Forest's patent drawing in figure 3.2 shows his audion connected in what amounts to a radio receiver with one stage of radio-frequency amplification. In fact, if you replace the audion (and its associated filament and plate-circuit batteries) by a simple diode, you get a radio receiver with no amplification, which will work with no extra external power supply at all, except, of course, the

No. 879,532. PATENTED FEB. 18, 1908.
L. DE FOREST.
SPACE TELEGRAPHY.
APPLICATION FILED JAN. 29, 1907.

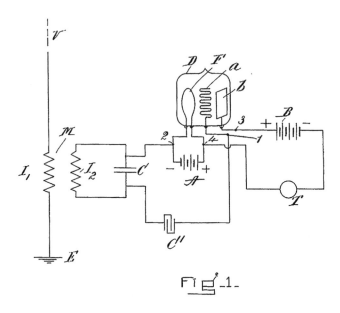

FIGURE 3.2. The principal figure from the De Forest patent for the audion, which also became known as the triode, or three-element vacuum tube. What is shown is actually a radio receiver with a triode detector and amplifier stage. The critical new, third, element in the "evacuated vessel" D is the "grid-shaped member a," now simply called the "grid." It controls the flow of electrons from the filament F to the plate b. The antenna V and ground E (earth) is at the left, feeding the radio-frequency transformer I_1–I_2 that transfers the signal to what we would today call the "LC tuned circuit I_2–C'." Finally, the output (plate) circuit contains T, a "telephone receiver," which in the early days of radio would usually be headphones. (After De Forest (1908).)

radio-frequency energy coming from radio stations. In early days, diodes were built by mounting a crystal (galena works well) so that it could be touched by a delicate pointed wire (called a "cat's whisker"), and millions of people in the early twentieth century heard their first radio broadcasts on such elegantly simple "crystal sets," truly minimal radios.

Returning to De Forest's patent, he did explain how his device works:[7]

> I have determined experimentally that the presence of the conducting member *a*, which as stated before may be grid-shaped, increases the sensitiveness of the oscillation detector and, inasmuch as the explanation of this phenomenon is exceedingly complex and at best would be merely tentative, I do not deem it necessary herein to enter into a detailed statement of what I believe to be the probable explanation.

Perhaps he thought it imprudent to explain too much, or maybe he was genuinely hazy about it. Today the explanation of how a triode works, with the hindsight of more than a century of scientific progress, is simple: When a negative voltage is applied to the grid, the electrons leaving the filament are repelled by the field around the grid and are blocked from reaching the anode. On the other hand, when positive voltage is applied to the grid, the electrons are drawn to the anode, and they flow from the filament to the anode. Hence the term *valve*: the grid acts very much like the handle of a water faucet in controlling the current through the tube.

We aren't used to seeing vacuum tubes today, but for two generations they were household items, essential components of every radio and television set. The tubes burned out about as frequently as lightbulbs, and many corner drugstores had tube testers so their customers could identify faulty tubes and replace them themselves. Tube designations like 6SN7 (a popular twin triode, two triodes in one envelope) or 6SJ7 (a *pentode*, having five elements) were as much a part of the popular vocabulary as the "8 megapixel" display screen, or the "16 GB" hard drive are today. Figure 3.3 shows a few examples over a 40-year span, illustrating one of the points we are aiming at in part I: they are all more or less (within a factor of five) the same size.

De Forest's audion was a breakthrough for a reason that had little to do directly with computers. As mentioned above, the tube could be used to *amplify* electrical signals, and that makes all the difference in the world when trying to capture a radio

FIGURE 3.3. Life before the transistor: six vacuum tubes, representative of the 1920s through the 1960s. From left to right: Cunningham CX345, 32 (GE), VR-105 (Hytron, also 0C3-A), 6SN7 (Sylvania), 6BQ7A/6BZ7/6BS8 (RCA), 5636 (Sylvania, imprinted "Engineering Sample," a subminiature). The Cunningham is $5\frac{1}{8}$ inches top to bottom. (From the author's garage.)

signal—which is what a lot of people were trying to do for the first time in the early twentieth century.

3.8 The Vacuum Tube as Valve

We're interested in vacuum tubes at this point not because they can amplify signals but because they can be used to build computers, where they play the role of controlled switches—really valves in a literal sense. Beyond that, we need to explain why the valve is really all you need to build a computer.

To this end let us consider how a vacuum tube can manipulate information in binary form. The essential point is that the presence of an input signal to the grid (in the form of a negative voltage) turns off the flow of current through the tube, whereas the absence of a negative voltage signal to the grid current allows the flow of current through the tube. That is, the flow of current is controlled by the grid voltage, just as the flow of water through a faucet is controlled by the handle. Notice also that if no voltage is applied to the plate of the tube, no current will flow in any event. After all, no water will ever flow if the sink is not connected to a water supply.

Think of the voltage applied to the plate as the INPUT logical value and the voltage app-lied to the grid as the CONTROL. The current through the tube gives us the OUTPUT signal; we

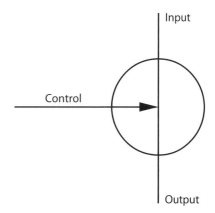

FIGURE 3.4. Symbol for an abstract, idealized valve. (After Schaffer (1988), p. 5.)

get this in the form of a voltage by tapping a resistor in the path of this current flow. We can thus arrange things so the following is true: the OUTPUT is ON when, and only when, the INPUT is ON and the CONTROL is OFF. That's all we need to abstract the essence of the valve. Schaffer uses the simple symbol shown in figure 3.4.

In the case of a vacuum-tube valve, we need to arrange some circuit details so that we can use the output voltage obtained from the plate circuit as an appropriate control signal. This means we need to assign the value ON to a negative grid voltage, which turns off the tube current, as described above; and we need to assign the value OFF to a zero grid voltage, which allows current to flow through the tube. Similarly, we need to ensure that the tapped voltages in the plate circuit correspond to appropriate OUTPUT signals; that is, the tapped voltages in the plate circuit need to correspond to ON and OFF signals at a grid, so the OUTPUT of one stage acts as CONTROL for succeeding stages.

These electrical arrangements should be clear to readers who have some familiarity with electrical circuits, but may be un-intelligible to those who do not. To end this chapter, we describe other ways that valves can be built, using sliding cams, electro-magnets, air flow, or semiconductors, and that should make it

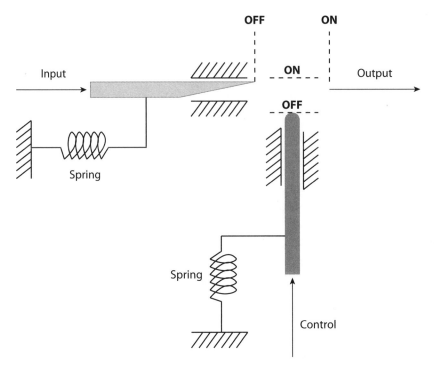

FIGURE 3.5. My attempt at a purely mechanical valve. The input rod reaches the ON position at the output if and only if the input is ON and the control is OFF. If the input rod is ON and the control rod is pushed from OFF to ON, the input rod slides back to the OFF position.

clear just what we're after here: there are many ways to build devices that can function as valves, some better than others in different situations, but we can think of any kind of valve as the basic building block of a digital computer. In principle, therefore, we can build computers that operate with air, water, or mechanical parts—instead of electricity. Figure 3.5 shows a valve that uses sliding cams, as mentioned above. In theory we could build a completely mechanical computer out of these—but I wouldn't want to try.

Before discussing more practical alternatives to the vacuum tube, we need to fill in a couple of missing pieces. We'll show that, using valves, we can carry out all the logical operations we need, build memory, and provide the clocking necessary to coordinate the logic gates and memory.

3.9 The Rest of Logic

Modern digital computers are built up in hierarchical layers, and we'll see now what the next layer would be if we started with nothing but valves. This is not the only way to organize things, but it's one way.

Starting with a box full of valves as our fundamental building block, the next step is to put together three kinds of gates that represent three fundamental operations of logic: the NOT gate, AND gate, and OR gate. The NOT gate has one input and one output, the AND and OR gates have two inputs and one output. The way they need to work corresponds naturally to usage in everyday language: If X is a two-valued (binary) digital value, NOT X is TRUE when X is FALSE and vice versa. If Y is another signal, X AND Y is TRUE when and only when both X and Y are TRUE. X OR Y is TRUE when and only when X is TRUE or Y is TRUE, or both.

We can build a NOT gate with just one valve. Think of the *control* line of the valve as the gate's *input*. Think of the valve's output as, simply, the gate's output. And then turn the valve's *input* line permanently ON.

If this is confusing, think about it this way: Put the valve in a box so we can't see what's inside, and think of the box itself as a gate with one input and one output.[8] Inside the box, where we can't see what's going on, connect the gate's input to the valve's control line, the gate's output to the valve's output line, and a permanent TRUE (with electrical parts, a battery) to the valve's input line.

If you like things more formal and algebraic, here's a third way to think about it. The valve is defined by the relationship output = (NOT control) AND input. If input is always TRUE, then output = NOT control, a NOT gate.

We build an AND gate using a NOT gate and another valve. Before connecting the new valve's control line, pass it through a NOT gate (constructed as above). Since the new valve's output is determined by the rule output = (NOT control) AND input, and since we are now applying NOT(NOT control), which is the same as control, the output of the new valve is output = control AND input, an AND gate.

Constructing an OR gate is now easy. Just observe that X OR Y is equivalent to NOT ((NOT X) AND (NOT Y)). That is, X OR Y is TRUE when and only when it's not true that both are false.[9] So we can build an OR gate with three NOT gates and an AND gate. As noted above, this is not the only way to do things. It may not even be a good way, but it's one way, and we only want to demonstrate the principle.

This kind of argument is typical of the way computer scientists think, and the layer-after-layer construction process can be continued from the bottom up to build instruction sets, sequencing circuits (so one instruction is followed after another), memory hierarchies (so data can be stored and retrieved), and so on, up to your browser, laptop, smartphone, and beyond. That would be fun to do, but it's done in many introductory computer books, and we won't do it here.

3.10 Clocks and Doorbells

If we let each gate go its own way in an interconnected collection of, typically, a few billion, there would be no control over the order in which the effects of signals would propagate from a designated input, through a particular set of gates, to a designated output. Chaos would reign. For this reason, digital circuits are usually provided with special signals that synchronize the gates, telling each one when to produce its new output from its most recent inputs, and then refreshing its inputs from the gates that feed it. In this way the logical steps performed by the gates march to a common drummer, called the *clock*. The term *clock speed* enters into the common specification of the computers we use every day and has become almost a household term. The chips in the computer I'm typing on have a clock speed of 3.4 GHz, which means the gates on the chips determine their outputs from their inputs 3.4 billion times a second.

One way to make a clock signal is to mimic the way an old-fashioned doorbell works: An electromagnet, which is just a coil of wire around an iron core, produces a magnetic field when current is sent through the coil. This attracts an iron clapper, which hits a bell. At the same time, the clapper draws apart a contact so the current in the coil is turned off. A spring then returns the clapper to its original position, turning the current back on, and starting the process all over again. In this way the clapper keeps hitting the bell, producing the familiar sound of a ringing doorbell. If you're too young to be familiar with this kind of doorbell, a buzzer works the same way.

If you look at the doorbell as a logical device, when it's ON (circuit contact in the closed position), it moves to the OFF position (opening the contact) and vice versa. This is the physical implementation of a logical paradox: ON implies OFF and OFF implies ON. We can build this using gates by connecting the output of a NOT gate to its input. What happens if we do this is that the gate alternates between its output being ON and OFF, the period of oscillation being determined by the time it takes for a signal's round-trip from input to output to input. The physical manifestation of a contradiction is thus a perpetual vacillation, or oscillation, which, in fact, can be used as a measure of time to "clock" computer logic.

3.11 Memory

The final missing piece is memory, another common specification of today's computers. Note again that what we describe next is not the only way to do things, but it illustrates the idea.[10]

In this case connect *two* NOT gates in tandem and return the output of the second NOT gate to the input of the first. It's now easy to see that this double NOT gate has two stable, consistent states: the first gate can be ON, the second OFF, or vice versa. In each case the second gate's output returns a value to the first gate's input that is consistent with its output, and the two-gate pair will maintain its state until it is forced into the opposite state. We won't discuss how to switch the two-gate pair from one of its states to the other. It is enough to say, "Ah, here is a circuit that remembers. We can use it to store information."

As promised, we have logic, clocks, memory, all built from valves. Time to look at a few nonelectronic kinds of valves before we return to the middle of the twentieth century and the next reason for the digital victory.

3.12 Other Ways to Build Valves

The vacuum tube, which I encourage you to think of as an electronic valve, changed the culture dramatically for the first half of the twentieth century. These warm and glowing little bulbs did so in two ways: the first as an analog device, the second digital. The analog applications were typically found in the radio and television sets (and even analog computers) of the appropriate era, the digital in what we now consider early digital computers. Both roles of the vacuum tube were, in the 1950s, assumed by the transistor, these days packed so tightly onto a silicon chip that the individual transistors are too small to see without a microscope.

It's important to understand how it happens that the vacuum-tube valve can function either as an analog or a digital device. In the first instance, the tube is operated in a range where changing the control voltage a little bit changes the output voltage a little bit, but over a proportionately wider range. We say in this case that the tube is operating in the *linear* range, because the output is proportional to the input. This is the role that vacuum tubes play in audio and video amplifiers, as well as oscillators if feedback is used.

In digital applications of vacuum tubes and, later, transistors, valves are used as gates, meaning that the input and output signals are *standardized*, as discussed above. That is, the signals are, at any one time, carefully kept very close to one of only two allowed values. This is decidedly not operation in the linear range, because the output voltage switches between two values, depending on, but not proportional to, the input and control voltages. More concretely, when the input and control signals are each either OFF or ON, the output is OFF or ON (depending on what kind of gate we're talking about).

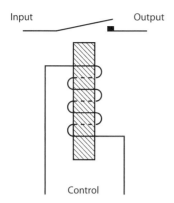

FIGURE 3.6. (Left) Diagram of an electromechanical valve, commonly called a *relay*. When the electromagnet is energized by current flow (CONTROL), its magnetic field pulls the contacts together (electrically connecting INPUT and OUTPUT). (Right) Photo of a modern relay, Omron model LY2F. The electromagnet is white, and the contacts are in the upper right of the plastic case.

The electromagnetic relay

Not all valves are as versatile as the vacuum tube and the transistor. The electromagnetic *relay*, for example, cannot be operated in a linear range. Its contacts are either open or closed, and therefore there is no intermediate range within which the output can be proportional to the input. It is strictly digital and never analog. Figure 3.6 shows a diagram of how it works. When current, as a control signal, is passed through the electromagnet, the contacts are drawn closed, which allows the flow of current in the output circuit. When the control signal is ON, the input is whatever the output is (OFF or ON); but if the control signal is OFF, the output is OFF regardless of what the input is. In one sense, the relay is a more primitive device than a vacuum tube, where the switch is controlled by a grid that influences invisible electrons flying in a near vacuum, and the first electromechanical relay predates the vacuum-tube valve by about 70 years.

Notice that in this particular example of a relay, the terminology is backward. When the control circuit is OFF (not energized), it disconnects the input and output, whereas in the vacuum tube, for example, the control (grid) being ON has that result. This is a trivial problem. We can simply call the control "OFF" when it's energized, or we could arrange the relay so the electromagnet pulls apart a connection that is normally closed. In fact, relays come in two flavors, "normally open" and "normally closed." The essential feature of a valve is that the connection between input and output can be *controlled by the output of some other valve.*

It is possible to build digital computers using relays. In fact, Konrad Zuse's machine Z3 used electromechanical relays, not vacuum tubes, and it can be argued that it was the first general-purpose, program-controlled digital computer, becoming operational on May 12, 1941.[11] The issue of priority is, however, somewhat controversial, for two reasons. First, the notion of a general-purpose, stored-program computer evolved gradually in several places and over a period of a decade or so. Second, Zuse's Z3 used a program that was external to the computing machine itself, and storing the program *internally*, in the same way as data, is considered crucial by many computer historians. Thus, Lavington dismisses Konrad Zuse's Z3 as one of a few "tentative pointers in the right direction,"[12] whereas F. K. Bauer, in his foreword to Zuse (1993), suggests the following epitaph for Zuse's tomb:

> Creator of the first fully automated, program-controlled and freely programmable computer using binary floating-point calculation. It was operational in 1941.

The usual account of computer development in the period immediately following World War II focuses on the US and Great Britain. The victors get to write history, and to Lavington, Zuse's Z3 is naturally an irritating example of German, if not Nazi, genius. Of course, the truth probably lies somewhere between the two pictures.[13]

Zuse's Z3 used 600 relays in the arithmetic unit and 1400 in the memory unit. Building a reasonably large memory was always a problem for relay machines; the Z3 had a storage capacity of only 64 words. Its program was, as we mentioned above, in some sense external to the machine, being stored and read on 8-track punched tape; the instructions were 8 bits each. Zuse cites the speed at "3 seconds for multiplication, division or taking the

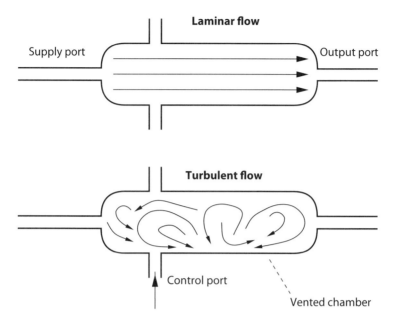

FIGURE 3.7. A valve using fluidics. With no stream applied to the control port, the horizontal flow from the supply port (INPUT) to the outlet (OUTPUT) is laminar. When a control stream is applied to the control port (CONTROL), the horizontal flow becomes turbulent and is vented, and the supply stream doesn't reach the output port. (After Markland and Boucher (1971), p. 11.)

square root."[14] Relays were unreliable and very slow compared with vacuum tubes, and it is a testament to Zuse's genius and persistence that he was able to build working relay computers, especially in wartime Berlin. The postwar efforts in the US and Great Britain used vacuum tubes almost exclusively.

The fluidic valve

The *fluidic* valve is sketched in figure 3.7. Its operation couldn't be simpler: A horizontal stream of a fluid (like air) is made to flow from an input supply port to an output port in a laminar flow. That is, the flow is straight and smooth, with flow lines that are nearly parallel. A control port makes it possible to inject a control stream of fluid at right angles to the laminar flow, which disrupts it and prevents it from exiting the output port. Instead, it is vented from the chamber. Thus, the device functions as a valve in the sense that there is an output signal only when the control signal is OFF and the supply port is ON, exactly what is needed to make a valve.

Fluidic means using the interaction of streams of fluid to process information. The "fluid" can be air, water, or any other fluid. Of course, air has the advantage of being harmless, so that unwanted streams can be vented with no garbage-disposal problem. Fluidic technology reached a high level of sophistication and peaked in the early 1970s, but was never a serious contender for replacing electronics because fluid switching is just too slow. It did, and still does, have niche application areas, because logic circuits can be made with no moving parts (not counting the fluid itself) and because fluidic circuits can work in hostile environments. They are immune to high temperatures and radiation, for example. Like the relay, the fluidic valve does not have a linear range of operation.

The transistor

The transistor started out in this world as a variation of the crystal set's cat's whisker, called the *point-contact transistor*, but was soon replaced by various forms of *junction transistors*. The details don't concern us now, but the device is an electronic valve, performing the same operation as the vacuum tube. Figure 3.8 shows a very simplified sketch of a field-effect transistor (FET), meant only to show its valve-like structure. In this case voltage applied to the gate (CONTROL) creates an electric field in the channel that controls the flow of current between the source (IN) and drain (OUT). We discuss how this works in a little more detail in chapter 4.

The general idea is the same, but the transistor has an all-important advantage over the vacuum tube. To make a vacuum tube, you need a vacuum and a tube. Then you need to heat the filament red-hot to boil off electrons. All this has to be assembled and the air removed from the tube; and when it operates, it takes up a fair amount of space, uses a fair amount of energy, and produces a corresponding amount of heat. A typical vacuum tube is a few inches high and uses on the order of a watt to heat its filament. That may not sound so bad, but a computer that

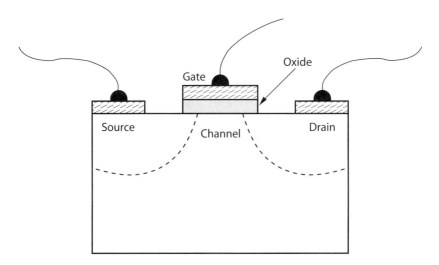

FIGURE 3.8. Idealized diagram of a field-effect transistor. The electric field in the channel caused by the voltage applied to the gate (CONTROL) either blocks or allows current flow from the source (INPUT) to the drain (OUTPUT). The charge carriers in the semiconductor body of the transistor are either electrons or the absence of electrons (holes). (After Roy and Asenov (2005).)

used the billion of them required for a very obsolete computer would burn a gigawatt, not including the cooling fans, and fill a large building—just for the tubes. And by the way, those red-hot filaments burn out the way lightbulbs burn out, and keeping a billion of them going at once would keep you on the run.

In a transistor, electrons (or the absence of electrons, holes) travel through a semiconductor instead of a vacuum. This means that loose electrons are wandering through the material, there to do our bidding, with no need for a delicate filament that glows red and no need for a vacuum or a tube to envelope it. Transistors therefore need much less power to operate, they stay a lot cooler than vacuum tubes, they can be made really tiny, and many more of them can be packed into a very small space—because there is less heat to get rid of. Adequate cooling is always an important consideration in building electronics, and we must always take care that our machines do not get hot enough to melt. For these reasons transistors completely revolutionized electronics in the post–World War II era. Radios, television sets, computers, and every other kind of electronic gadget became much smaller, cooler, and cheaper. Transistorized portable radios (now an obsolete phrase) were called "transistors," a forgotten synecdoche.

A very natural question now arises: Just what limits the number of transistors that we can ultimately pack into a given space? To answer this we need a little more fundamental physics, which we take a look at next.

4 Consequential Physics

4.1 When Physics Became Discrete

As the nineteenth century rolled over to the twentieth, fundamental physics was radically transformed. It was not an evolutionary flowering; it was an earthquake. The electron was discovered, the photon recognized as the particle of light, and the structure of the atom subsequently unveiled; the physics of the microscopic world became discrete. A few decades later, information processing experienced the same transformation from continuous to discrete.

Revolutionary change was in the air. After all, within the same few decades, Einstein gave us special and general relativity (changing space and time forever); Gödel gave us his incompleteness theorems (changing true and false forever); and Stravinsky gave us *The Rite of Spring* (changing music forever, as well as causing an uproar in 1913 Paris).

However, the connection between the development of quantum mechanics and the digital computer revolution is quite direct, and there is no need to invoke a vague notion like the zeitgeist to explain the change. It was the development of twentieth-century physics that rendered obsolete the glowing vacuum tubes of my childhood. Quantum mechanics explains how transistors work and reveals a source of absolutely unavoidable physical noise and granularity and, with that, the ultimate limitations of analog devices. It is fair to say that without quantum theory we would not be able to design and produce the tiny, highly dense semiconductor chips that we now depend on so much in our day-to-day lives.

Another, perhaps more surprising, reason for exploring physics in this book is that it makes possible a new kind of computer—the *quantum computer*. But that's getting ahead of ourselves; more about that later.

Let us pick up the story of quantum mechanics in the year 1900, when Max Planck explained blackbody radiation in an apparently ad hoc way. Blackbody radiation was presenting the physics establishment with a wicked problem. To explain the conceptual setup, think of a large oven with a small aperture. Any radiation that enters the aperture never comes out—it is a perfectly absorbing (black) window. But radiation leaves, and different amounts of energy will leave at different frequencies; the distribution of energy versus frequency is called the *spectrum* of the observed radiation (as in section 2.5). Physicists worked out that distribution, and at low frequencies all was well. There was good agreement between the predicted and the experimentally observed spectrum. But at high frequencies all hell broke loose: the predicted energy got more and more intense with frequency, and the classical, pre-quantum-mechanical theory at the time predicted that an infinite amount of energy would emerge from the aperture of the oven. This is certainly not in accord with observation; it's a catastrophe. The effect is, in fact, called the *ultraviolet catastrophe*, ultraviolet because it happens as a result of the total radiation at very high frequencies.[1]

Max Planck then pulled a rabbit out of a hat, albeit with a highly educated hand. He postulated that the energy transfers that take place in the oven can involve only *integer* multiples of a fundamental quantum of energy—in our terms, that the allowed values of energy are *discrete*. This assumption leads to a predicted spectrum of radiation leaving the aperture that is in excellent agreement with experiment. In a famous letter written much after the fact, Planck reports:[2]

Briefly summarized, what I did can be described as simply an act of desperation. By nature I am peacefully inclined and reject all doubtful adventures. But by then I had been wrestling unsuccessfully for six years (since 1894) with the problem of equilibrium between radiation and matter and I knew that this problem was of fundamental importance to physics; I also knew the formula that expresses the energy distribution in normal spectra. A theoretical interpretation therefore *had* to be found at any cost.

Planck announced his result on December 14, 1900, the date generally known as the birthday of quantum mechanics.[3] Energy had been made discrete.[4]

Five years later, Albert Einstein took another audacious step. The puzzling problems this time—and real progress in science is often stimulated by puzzling problems—concerned the *photoelectric effect*. When a light beam hits a piece of metal, electrons are knocked out of the metal's atoms. In a vacuum they can be collected, and this is one way to detect light. You can make an automatic door opener this way, using an old-fashioned vacuum-tube photoelectric cell (instead of a solid-state device). By the time Planck had proposed that energy comes in discrete packets, serious problems had been found with the classical (nineteenth-century) theory of the photoelectric effect.

One problem was this: If a beam of light is shone onto a metal, you can calculate the amount of time it takes for enough energy (per unit area) to be absorbed to knock out an electron. It can be a few seconds if the light is dim enough, and it would be impossible for any current to begin before this minimum amount of energy arrives. But what is observed is that some electrons get knocked out before that time is elapsed. This is difficult to explain if light is a continuous wave. But Einstein pointed out that if light itself arrives in discrete packets, what we have already called *photons*, the phenomenon is easy to explain. It takes only single photons here and there to begin dislodging electrons before the initial calculated time has elapsed.

Another serious problem had been observed with the classical theory of the photoelectric effect. There is a certain minimum frequency of light below which no electrons are dislodged at all. Furthermore, above this threshold the maximum energy of the dislodged electrons doesn't depend on the intensity of the light beam but only on the frequency of the light. If the intensity of the illumination is increased, more electrons are produced, but their maximum energy is the same. Again, Einstein explained this with his discrete theory of light. If light is composed of photons and those photons have an amount of energy that depends on their frequency, then no electrons will be dislodged if the photons are not energetic enough; above that threshold each photon will give its quantum of energy to an electron. Increasing the intensity of the light only increases the *number* of photons, and hence the number of electrons produced—but not each electron's energy.

Einstein put it this way:[5]

> The energy of a light ray spreading out from a point source is not continuously distributed over an increasing space but consists of a finite number of energy quanta which are localized at points in space, which move without dividing, and which can only be produced and absorbed as complete units.

Now light had been made discrete.

It took almost 20 years for the final step to be taken. Light was thought to be a wave but can behave like particles. So why can't a particle—like an electron—behave like a wave? Louis de Broglie proposed this in 1924, and the idea was confirmed when diffraction patterns were produced by scattering electrons off a nickel target. Since then, the wave-particle duality of other particles, as well as atoms, and even molecules, has been demonstrated experimentally. Thus, energy and light had been made discrete, and matter had been "wavified." Everything in the world had become both a particle and a wave. This poses a very pretty puzzle, which we are now in a position to discuss in terms of quantum mechanics. The ideas will serve us well when we turn our attention to the twenty-first-century world of information.

4.2 The Absolute Size of Things

It follows from the discretization of matter and energy that the words *big* and *small* are not just relative terms—they have an absolute significance. They define *scales*. For example, we can talk about the two extreme scales of size in the universe, and the scale between the two, the scale at which we live. Call these the *subatomic*, *everyday*, and *astronomical* scales. At the everyday scale, we consider distances that are on the order of a single meter. At the low end, when we consider particles like electrons and protons, we deal with things that are about 15 orders of magnitude smaller, where we use the term *order of magnitude* to mean a factor of 10, which is usual scientific parlance. And at the astronomical scale, a convenient length is 10 petameters (10^{16} m), roughly a light-year.

So it turns out that we human beings are just about in the middle of the range of size scales, logarithmically speaking. This explains why the classical physics that reached maturity at the end of the nineteenth century worked so well at the scale of everyday life. It was when it began to be tested at the subatomic and astronomical scales that serious cracks developed in the foundations—with photons and electrons at the subatomic scale and the very odd absolute speed of light and absence of the ether at the astronomical scale.

Where, then, can we find a good standard of absolute size, a meter stick for the universe? Physicists call the numbers that determine the scale of things *fundamental constants*. They are sewn into the fabric of the universe, unalterable, and, as far as we can tell, the same everywhere in it. The speed of light, always denoted by c, is one example, probably the most well known.

The fundamental constant that determines subatomic scale lies in the work of Max Planck that we mentioned above, as well as in the foundations of quantum mechanics worked out by Werner Heisenberg and his colleagues in the late 1920s. Recall that Planck observed that the energy distribution of the radiation emerging from an aperture of an oven—blackbody radiation— could be explained by energy transfers taking place only in discrete packets. Specifically, he proposed that the size of the packets was a constant h times the frequency of the radiation. This constant is now called *Planck's constant*, and it plays a critical role, along with the speed of light, c, in determining the scale of things.

4.3 The Heisenberg Uncertainty Principle

For a particle to qualify as a particle, it must have a well-defined position. We would like to be able to say that a particular particle is precisely at such-and-such a place. We may also want to specify its velocity. That is, we would like to say that the particle is at such-and-such a position and is traveling at such-and-such a speed to the right, say. Only then would it be fair to claim that we are dealing with a bona fide particle—using the term *particle* as we do in everyday life. Quantum mechanics says that this is actually *impossible,* a result at the heart of the theory. It came as quite a shock, given our experience with everyday objects like baseballs. But remember that an electron is 14 orders of magnitude smaller than a baseball. The rules change when things get that small. In fact, the very notion of *size* becomes fuzzy at that scale, which is one of the principal messages of quantum mechanics.

To talk about the position of a particle, say, one must really describe how to measure position. No measurement is perfect, and no matter how we measure something like position, there is some uncertainty associated with the result. The same is true for velocities. The Heisenberg uncertainty principle states that the *product* of these two uncertainties can never be less than some very small number divided by twice the mass of the particle. The very small number, a universal constant, is Planck's constant (mentioned above) divided by 2π (for convenience), which is called the reduced Planck's constant, $\hbar$. What's important here is that $\hbar$ is small—*really* small, about 10^{-34}, or roughly 34 orders of magnitude less than unity.[6]

Suppose we want to measure the position of a baseball at some particular time, say, when it's crossing home plate on its way from the pitcher to the catcher. Suppose we then take a photograph with a high-speed shutter and narrow down its position to a fraction of a millimeter. The uncertainty principle states that if we want to know the *velocity* of the ball at the same instant, we can't measure it more accurately than some very small number divided by the mass of the ball.[7] But that result is still very, very small—much smaller than we could ever measure with the best available camera. In everyday life, the uncertainty principle doesn't really tie our hands, and it allows us to treat the objects around us in the way we do. The evolution of our species over hundreds of millions of years has ensured that the ideas we have of position and velocity work in practice in our macroscopic world.

To take a concrete example, it is possible to measure the position of a paper clip, which has a mass of about 1 gram, to within a billionth of a meter, and simultaneously measure its velocity to within a billionth of a meter per century, and still respect Heisenberg's uncertainty principle (with a large margin to spare).[8]

In the world of the electron, however, the picture changes dramatically. The mass of an electron is about 10^{-27} times that of a paper clip, and since it's so small we might imagine that we want to measure its position to within an atomic diameter, which is roughly 10^{-10} m. The uncertainty principle then limits the precision of a velocity measurement of the electron to just about half a million meters per second or, in terms more familiar to motorists in the US and Great Britain, about a million miles per hour! In other words, if we want to say that a particular electron is in a certain atom, we can say essentially nothing at all about its velocity.

It is worth emphasizing the fact that Heisenberg's uncertainty principle, at the heart of quantum mechanics, is not any sort of practical limitation that can be overcome by buying better equipment or being more careful in the laboratory. It is a fundamental limit on what we can know about things around us. Given that quantum mechanics has been confirmed to great precision in many ways by countless experiments over the last century, it seems very likely that it is a limit that will never be beaten in this universe.

4.4 Explaining Wave-Particle Duality

In the next chapter we descend into the world of the very small, in search of the smallest that computers can get. But before we continue along those lines, we discuss two more extremely important aspects of physics at very small scales. First, consider briefly the apparent paradox of wave-particle duality. How is it that something like an electron, for example, can behave very much like a particle in one situation and very much like a wave in another? A full discussion of the question would require much more detail about the measurement process in quantum mechanics and would take us far afield. But the uncertainty principle can give us some very nice intuition.

To repeat (and it is worth repeating), the uncertainty principle states that the *product* of the uncertainty we have about the position of an electron, say, and its velocity can never be smaller than some definite, fixed, God-given number. Suppose we narrow down the uncertainty in position to some very small amount. The smaller we narrow down our knowledge of the electron's position, the less we can know about its velocity. If we continue to narrow down our uncertainty about the electron's position—with some measurement process that we needn't discuss—to the point when we can claim it has a definite position, it is reasonable to regard it as a *particle*. In the world of quantum mechanics, particles have definite positions, waves do not. At the other extreme, if we try to measure the velocity of an electron to greater and greater precision, we must of necessity, by the uncertainty principle, be left knowing less and less about its position. In this latter case it is legitimate to regard the electron as a wave (in the world of quantum mechanics) and definitely not a particle. The uncertainty principle provides at least a plausible justification for the strange behavior we observe: electrons sometimes hit metal plates like microscopic baseballs, and sometimes they interfere with each other and bend around corners like waves on the surface of a pond.

4.5 The Pauli Exclusion Principle

In addition to wave-particle duality, we need one more principle of quantum mechanics to understand today's computer chips, the *Pauli exclusion principle*. A reminder before we go on: the Pauli principle is couched in terms of "particles." But we now know that particles are really also partly waves, and waves are really also partly particles. So, for example, when below we speak of "clouds of electrons," we are picturing electrons in the very cramped quarters of atomic orbitals, and they are behaving like both particles and waves.

We see in the next chapter how Heisenberg's principle sets the ultimate limit on how much miniaturized circuitry can be squeezed onto a semiconductor computer chip. The Pauli exclusion principle explains how semiconductors work in the first place. In fact, without exaggeration, the exclusion principle makes our whole world possible. Without it, we would not only be at a loss to explain how semiconductors work, but we could not even explain why we have the elements that we have, and how most everything else on earth is put together. The principle explains why the elements can be arranged in neat rows and columns in the periodic table, why neon is inert, why oxygen is so eager to combine with other elements, how proteins are built, and so on. To get an idea of what the exclusion principle tells us, we need a little more background about fundamental particles.

In this book we need to consider only two different kinds of particles: photons and electrons. As you know, there are many other fundamental particles. The two others most often mentioned are protons and neutrons, which usually mind their own business, safely nestled in the nuclei of atoms. When they are knocked loose, the "CAUTION: RADIOACTIVITY!" signs go up. High energies are required to liberate them from their cozy nuclear homes, and we need to stay out of their way. But the electrons in atoms are more loosely packed in clouds around the positively charged nuclei, and the electrons that figure in the operation of semiconductors carry much less energy per particle. There is no danger in getting hit by a stray electron or two.[9] It is the behavior of these electron clouds around nuclei that determines how all of chemistry works. Some electrons are much less tightly bound than others to their parent nuclei, and these electrons can easily go wandering off, accounting for the conduction of what we call "electricity" in metals (which is nothing more than the flow of electrons). Photons are even more free—in fact, they cannot stay still at all but fly around at a phenomenally high speed.

Photons account for visible light, but they also account for many other types of radiation, the difference being only in the photon's wavelength and energy. As we've noted before, a photon's energy and frequency are related in a very simple way. The energy of a photon is proportional to its frequency, and the constant of proportionality is Planck's constant. Ultraviolet light has energy above that of visible light, and infrared below. X-ray and gamma-ray photons have even higher energy, radio waves even lower. But all these particles (or waves, as you wish) are described by the same mathematical constructs; they're all photons, all wave-particles. We don't have to worry here about the peculiar properties of photons, but they *are* quite peculiar. For example, in a vacuum, they always appear to be traveling at the same speed, no matter how fast you, as an observer, are traveling. You can never catch up with a photon. That's pretty peculiar in itself.

Electrons also have energy associated with them. But whereas the energy of a photon is determined entirely by its frequency, the energy of an electron is of a more conventional sort and is determined by the amount of work that has been expended to get it where it is. For example, the electrons in the inner shell of an atom, closest to the nucleus, have a low energy. They can be kicked to a shell farther from the nucleus by some force, and when they are in the outermost shell of an atom, they can be kicked free. But all this kicking takes energy, which the electron carries with it.

Recall in all this discussion that energy comes in discrete packets (is *quantized*). Energy is also *conserved*. That is, the total amount in a system we're considering must remain the same; it can neither be created nor destroyed. It may happen, for example, that a photon hits an electron to kick it up to a higher energy state. The photon in such cases either gets absorbed or, if there is energy left over, bounces off at a lower energy. Conversely, it may happen that an electron in a high energy state falls back to a lower energy state. What happens then to balance the books is that the extra energy is ejected as a photon, with just the right frequency to account for the new photon's energy.

Now, electrons and photons are exemplars of two fundamentally different types of particles, *fermions* and *bosons*, and all particles are either one or the other. Electrons are fermions and photons are bosons. Protons and neutrons are fermions. The Higgs boson, which made news when its existence was confirmed only in 2013, is, well, a boson. The point we're leading up to is that fermions are constrained to behave in a way that bosons are not. Fermions must obey the Pauli exclusion principle, and it is this principle that is essential in understanding semiconductors, the wonderful materials at the heart of our computers.

To put it simply, the Pauli exclusion principle states that fermions are extremely antisocial. An electron in a given quantum-mechanical state will never tolerate another electron in the same system in exactly the same state. (In contrast, bosons are gregarious and are not bound to obey the Pauli exclusion principle.) To understand what this entails, we should explain exactly what is meant by the term *quantum-mechanical state*, but that would require more space and mathematics than I am allowing myself here. At this point, it is sufficient to know that the *state* of an electron bound to a nucleus in an element is specified by four numbers, one of which is called its *spin*, which can take on only two possible values, $+\frac{1}{2}$ and $-\frac{1}{2}$. The exclusion principle then states that *no two electrons in an atom can have exactly the same set of these four numbers*.

4.6 Atomic Physics

From this simple rule, we can at least see how the first few elements in the periodic table are put together.[10] The hydrogen atom is the simplest and is the easiest to picture. A proton, with a charge of $+e$, holds a single electron, with the charge $-e$, in an orbit surrounding it. The use of the term *orbit* is conventional; we know that an electron is really part wave and part particle, and the idea of it actually *orbiting* a proton, as the Moon orbits the Earth, is naïve.

The next simplest atom is helium, with two protons in its nucleus,[11] and consequently a charge of $+2e$. To balance that charge, two electrons will be attracted to the nucleus, and they can fit in the same "orbit" without violating the exclusion principle because they will have different spins. The electrons around the nucleus of heavier atoms are arranged in *shells* that (generally) fill up with electrons from the inside out. The first shell is complete with two electrons, in the sense that two electrons in the first shell are quite stably bound to the atom and are very reluctant to go wandering off. As a result, helium is not interested in sharing its electrons with other elements, and under normal conditions does not form compounds.

The outermost shell of electrons of any atom is called its *valence* shell, and the electrons in that shell are called *valence electrons*. For example, the second shell can hold up to eight electrons, and the element with full first and second shells is neon, with a total of ten protons and ten electrons. Elements like helium and neon, with complete valence shells, are all gases, and they're called *noble gases* because of their elitist attitude.

The next heaviest element after helium—lithium—has three protons in its nucleus and three electrons: two in its innermost shell and one more electron to start its second, valence shell. The inside shell, with two electrons, can be regarded as permanently closed, but the one electron that starts the next shell is quite loosely bound to the rest of the atom and is only too happy to join another atom, leaving behind a charged ion, or even to go wandering off as part of an electric current. The element lithium is, because of that loosely bound electron, a metal and a good conductor of electricity.

The availability of valence electrons plays a crucial role in determining the electrical properties of a material.

4.7 Semiconductors

A semiconductor is (usually) a crystalline material composed of atoms of some basic material, usually silicon today, held in place by shared electrons. The silicon atom has four electrons in its outside shell, and, as usual for our purposes, it is only these valence electron shells that matter; the inside shells are all filled completely with all the electrons they can hold, and they will not be disturbed by the adventures of the electrons in the valence shell. If such a crystal structure is perfect (so that there are no "loose" electrons free to wander around the crystal) and the temperature is absolute zero (so that there is no thermal vibration that could knock electrons free), the crystal does not allow the conduction of any electricity—it is a perfect insulator, a rigid crystal with all its electrons held firmly in place, bonding adjacent atoms.[12]

But if the crystal is imperfect (which it always is), and the temperature is above absolute zero (which it always is), then there are at least *some* electrons free to move about the crystal lattice, which allows the flow of some electrical current. At reasonable temperatures, some electrons can jump out of their usual positions and wander through the crystal. Not only that, but when the electrons leave their usual place, they leave "holes," and those holes can, in effect, also travel through the crystal in the same way as do real particles. An electron can jump into a hole, leaving a hole where the electron had been previously, and in effect the hole moves as a "virtual particle," with a positive instead of negative charge.

To make valves (or what amounts to the same thing, transistors) from a semiconductor, we don't rely solely on the electrons freed by thermal vibration or natural faults in the crystal. The semiconductor crystal lattice is intentionally contaminated with a material called a *dopant*. Typically, one atom out of a few million silicon (say) atoms in a crystal lattice is replaced by an atom of another material, called the *dopant*. A tiny bit of doping can have a dramatic effect on the way the crystal conducts electricity.

Suppose then that we replace about one in a million silicon atoms in a crystal with arsenic atoms, an element that happens to have five (instead of four for silicon) electrons in its valence shell. The arsenic atom can sit in the place of a silicon atom, but there will be an electron left over. For practical purposes we can think of this electron as free to move about, almost as if it had been released from a filament in the vacuum of a vacuum tube. Notice that, although there are now free electrons in the crystal lattice, the net charge of the crystal is zero. For every free electron there is an arsenic atom that has a net positive charge of $+e$ because of its errant electron; it is called an *ionized donor* and can be considered fixed in the crystal lattice. A piece of silicon doped in this way is called *n-type* silicon.

In the same way, we can replace silicon atoms with, say, aluminum, which happens to have only three electrons in its valence shell. This creates holes in the lattice, which, as we described above, behave effectively like positive particles that are also free to move about the lattice. The dopant atom that has a negative charge because an extra electron has jumped into its valence shell is called an *ionized acceptor*, in analogy to the case above. A piece of silicon crystal so doped is called *p-type* silicon.

It is important to realize that two kinds of electrical conduction can take place in a doped semiconductor. In n-type silicon (say), there are free electrons, and their flow can constitute an electrical current, in the same way as it does in a real metal, where there is an even greater abundance of free electrons. We say in this case that the electrons are *charge carriers*. In p-type silicon, the holes can flow, effectively acting as particles that carry positive charge. The charge carriers in this case are holes. If we connect a battery across a piece of n-type silicon, it creates an electric field inside the silicon, and electrons flow under the influence of that field. In the same way, connecting a battery across a piece of p-type silicon results in a flow of holes in the silicon. In general, both electrons and holes are potential charge carriers. Also keep in mind that the (charged) ionized donor and acceptor atoms are sitting locked in place in the semiconductor lattice.

4.8 The P-N Junction

Something very interesting happens if we take a piece of n-type and p-type silicon and carefully join them face-to-face, forming what is called a *p-n junction*. Figure 4.1 shows a sketch of such a junction, with no voltage applied externally. There are lots of free electrons on the n-side of the junction, and these tend to wander randomly (diffuse) to the p-side, where they can jump into holes. Similarly, holes tend to diffuse from the p-side to the n-side. This redistribution of charge carriers creates a negatively charged wall on the p-side of the junction (because of the unbalanced acceptor ions) and a positively charged wall on the n-side (because of the unbalanced donor ions). The negatively charged wall on the p-side builds up until its repulsion of electrons prevents their further diffusion into the p-type silicon;[13] and the positively changed wall on the n-side prevents the further diffusion of holes. An equilibrium is then reached where there is a region around the junction with a shortage of charge carriers, called the *depletion region*.

Now consider what happens if we apply a battery across a p-n junction with its positive terminal connected to the p-type silicon and its negative terminal to the n-type silicon. The positive terminal pushes the holes toward the n-type silicon, the negative terminal pushes electrons, and the effect is to shrink the width of the depletion region. If the battery voltage exceeds some

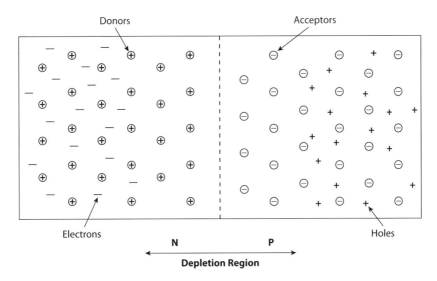

FIGURE 4.1. A p-n junction in equilibrium (with no applied voltage). To the left, n-type semiconductor with positively charged donor ions in the lattice indicated by "⊕" and electrons by "−". To the right, p-type semiconductor with negatively charged acceptor ions indicated by "⊖" and holes by "+". Electrons diffuse to the right, and holes to the left, across the junction (shown dashed), until the negatively charged wall of ions on the right and the positively charged wall of ions on the left prevent further diffusion. This leaves a depletion region around the junction with a shortage of charge carriers (electrons and holes). (After Bar-Lev (1993), p. 99.)

(generally small) threshold, current flows. Actually, electrons flow from left to right in the n-type silicon and jump into holes at the junction, those holes having arrived in their journey from right to left in the p-type silicon. In the figure, with the battery connected as assumed, electrons would be injected at the left and sucked out at the right (in effect, injecting holes).

However, if we connect the battery in the opposite way, things do not go well for conduction. The positive and negative battery terminals pull the charge carriers away from the junction, and the depletion region gets wider. No current flows. Thus, the p-n junction forms what is called a *diode*, which allows current to flow in only one direction.

The p-n junction encapsulates the magic of semiconductors. To make a vacuum-tube diode, we need to provide a vacuum where electrons can move freely (in one direction only), and a hot source of electrons. Now we have a way of creating a diode in a solid material without a hot filament. This is the point: we have replaced vacuum-tube technology with *solid-state electronics*.

4.9 The Transistor

It is now not hard to see how one can take advantage of the movement of electrons and holes (really the absence of electrons) in semiconductors to build a valve. There are many variations on the theme, all called *transistors*, but the idea is perhaps most clearly illustrated by the field-effect transistor, discussed in chapter 3 and diagrammed in figure 3.8. The *source* and *drain* in the figure are both of the same type of doped silicon, say, n-type, and they are separated by the *channel*, which is oppositely doped.[14] Without worrying for now about the third wire connected to the transistor, there are now two depletion regions, one where the channel meets the source and one where it meets the drain, and current will not flow between source and drain. The transistor, as a valve, is OFF.

Consider, to be concrete, an n-p-n transistor. The mobile electrons in the source and drain cannot flow through the channel, precisely because they face negatively charged ions at the depletion regions. The *gate* is the handle that can close the valve. It is a small piece of conducting material (a metal, say) that sits above the channel, but is insulated from it. If now we apply a *positive* charge to the gate, it creates an electric field in the channel that attracts electrons to the channel, even though it is not connected to it electrically; thus the name *field-effect* transistor. Suddenly, there are charge carriers in the channel, current can flow, and the transistor is ON.

Pauli's exclusion principle, which explains where the electrons are in a semiconductor crystal and where they can go, has led us to a solid-state valve. And, as we've seen, a valve is all we need to build any kind of computer we want.

If you plan to study electronics in some depth, there are many excellent books on the behavior of electrons in solids, at a progression of levels. They all use quantum mechanics, and the more advanced they are, the more quantum mechanics they use.

The atoms in a crystal are small, electrons are even smaller, and the important mechanisms cannot be understood at all without the science of the very small. These books are easy to find, and your choice will of course depend on your level of mathematics and physics. The book cited above, Bar-Lev (1993), is a thorough and fairly advanced text, not for the beginner, but it does contain much more detailed (and mathematical) analyses of the p-n junction and field-effect transistor discussed here, together with interesting descriptions of the integrated circuit technology used to produce our chips, without which life would be, well, a lot less interesting.

Incidentally, it may surprise you to know that it is actually possible to build vacuum tubes and transistors in your home workshop, with ordinary tools and materials. The advanced tinkerer (or armchair tinkerer) can find details of how to do this, along with expert advice and background, in the delightful book *Instruments of Amplification* by H. P. Friedrichs (2003).

4.10 Quantum Tunneling

We end this chapter by describing, very briefly, *quantum tunneling*, one more quantum phenomenon that plays an important role in our coming descent into the realm of very small electronics. Particles like electrons are charged, and when they move around in electrical fields, they behave very much like particles rolling around on a surface with hills and valleys. The field-effect transistor provides an example of this. When the gate is not charged, the electrons in the source cannot reach the drain, because the channel presents a *barrier* to their progress. The situation is analogous to a rolling ball encountering a brick wall. There is, however, an important difference. The electron is very small, and the laws that govern its motion are quantum mechanical, not classical. This has a counterintuitive consequence: electrons *can* penetrate barriers under certain circumstances, with a certain probability, if the barrier is not too thick.

When electrons, or any other particles, penetrate what seem like solid barriers, we say they *tunnel*. It's one more thing that can happen in the quantum-mechanical world of the very small that does not happen in our everyday experience. It means that the width of the channel in a field-effect transistor, or the gate in other kinds of transistors, can get so small—but no smaller. There comes a point in shrinking the size of transistors when electrons will tunnel from source to drain, and the valve no longer works as it should. In the next chapter we examine the miniaturization of electronics and its limits, and electron tunneling presents one such limit.

4.11 Speed

So far we've paid a lot of attention to the *size* of things but not the speed with which they operate. In the world of transistors, however, the two are very closely related. The reason stems from the fact that charge is stored in a semiconductor transistor, and it takes time for the charge that is built up in a depletion region, say, to change appropriately when the transistor switches from ON to OFF or OFF to ON. Whenever charge is stored, the place in which it is stored is called a *capacitor*, and the ratio of the charge that is stored to the voltage that is applied to the capacitor is called its *capacitance*. Thus, for a fixed voltage the charge stored in a capacitor is directly proportional to its capacitance.

Classically, there are three electrical components in a circuit, leaving aside so-called active elements like transistors or vacuum tubes: *resistors* (which impede the flow of electrons), *capacitors* (which store energy as charge), and *inductors* (which store energy as a magnetic field). If you dissect one of the beautiful radios I described that ruled the 1930s, you will find all three kinds of "passive" elements, each with two wires emerging from it, usually soldered in place. Today the three elements still play a central role in our understanding of electronic circuits, but the resistors, capacitors, and inductors are usually abstractions of our understanding of how microscopic devices like transistors work. The transistor, for example, can be thought of as behaving as if it is a combination of resistors and capacitors. The depletion region is one example of such a microscopic, conceptual capacitor.

In beginning physics courses, the capacitor is introduced ideally as a pair of flat metal plates, parallel to one another, and spaced a certain distance apart. One of the plates gets charged with an excess of electrons, and the other with a deficiency. For a given spacing between the plates, the capacity of such a *parallel-plate* capacitor is proportional to the area of the plates. By rough analogy, the capacity of something like the depletion region of a transistor shrinks directly with the size of the transistor.

It's not hard to see, then, that the time it takes for a capacitor to get charged is roughly proportional to its area, all other things (like resistance and voltage) being equal. The smaller we can make transistors, the faster they can switch, and hence the faster they can be clocked in a computer. That is why the dramatic shrinking of transistors that we are about to describe next was accompanied by a similarly dramatic increase in speed.

5   Your Computer Is a Photograph

5.1   Room at the Bottom

Richard Feynman, a hero to more than one generation of physicists and computer scientists, brought wide attention to the potential of the very small in a famous talk to the American Physical Society, on December 29, 1959. In that talk he anticipated, by a few decades, the now burgeoning field of nanotechnology. He had a way of getting directly to the heart of his subject, at the same time making it all seem obvious. It's simple, anyone could have seen it, and, as he says at one point after describing how the *Encyclopedia Brittanica* can fit on the head of a pin, "I don't know why this hasn't been done yet!"[1]

Not to take anything away from Feynman, who was talking about more general adventures in the nanoworld, something along those particular microphotographic lines *had* already been done, a generation earlier.[2] Emmanuel Goldberg, in 1925, produced photographic images with such a high density that 50 complete bibles would fit in a square inch.[3] At that time, in fact, there was something of a competition among producers of highly reduced images, and the "bible per square inch" had become a common measure of text density. Incidentally, the current record is claimed by a team at Israel's Technion,[4] who reported etching the equivalent of about 2500 bibles per square inch on a thin layer of gold, using an ion beam.[5]

Writing small seems to be, for some reason, a natural human obsession. The Assyrians produced cylindrical clay tablets 5000 years ago with writing so small that a magnifying glass is required to read them.[6] When in 1839 L. J. M. Daguerre produced the first photograph on an iodized silver plate developed with mercury vapor, the "Daguerrotype," J. B. Dancer made the first microphotograph, at a reduction ratio of 160:1—within the very same year.

An obvious advantage of such reduction is saving space, and the technique was put to use in 1870 when R. Dagron used microphotography to supply cargo for carrier pigeons during the siege of Paris.[7]

Besides the obvious advantage of saving space and weight, microphotographs have had some important applications in espionage. A "microdot" is a microphotograph the size of a typographic dot, such as the one that ends this sentence. Microdots can be sprinkled throughout a letter, for example, in place of ordinary printed dots, so that information can be transmitted undetected by unknowing eyes. They can then be read with a microscope by a recipient who knows where to look. Aficionados of the spy-novel genre know this as a standard trick, in every good spy's repertoire.

I should point out an important distinction: there is a considerable difference between producing an image of a page in a book and just recording the digital information, the bare minimum for reading content. The best in the former category shows images of letters and preserves things like the font, photos, and smudges from candy bars. It's what a scanner produces. Images in the latter category usually use just one byte (1B, or 8 bits) per character, so there are 64 possible characters. We refer to the former as *text images*, or text in *image form*, and the latter as *digital text*, or text in *digital form*.

The Hebrew bible has about 1.2 million letters. Let's say that's 1.2 MB. A 128 GB flash drive, a common consumer item as this is written, therefore holds about 100,000 Hebrew bibles in digital form, and its memory chip is on the order of a square inch in area. We can then say that such a flash drive stores 100,000 bibles per square inch, but keep in mind that this kind of bible is not the same as the page images that were written at Technion on a speck the size of a grain of sugar. We are interested more in imaging in what follows, because we want to get back to valves and the problem of making small computers.

Feynman (1960) was well aware of the distinction between text images and text in digital form. He also went a step beyond the writing (and reading) of digital text on a flat surface, and considered the storage of such text in a three-dimensional chunk of matter. He argued that, to store a bit, we would need at a minimum a cube of matter 5 atoms on a side, or, in our customary style of rough estimates, about 100 atoms per bit. This factor of 100 over the ON/OFF, there-not-there, bit provides some redundancy to take care of some possible loss of information. His conclusion is that "all of the information that man has carefully accumulated in all the books in the world can be written . . . in a cube of material one two-hundredth of an inch wide—which is the barest piece of dust that can be made out by the human eye." Even allowing for the explosion in information that has occurred since Feynman's talk in 1959, as he says in his title, there's *plenty* of room at the bottom.

It is worth mentioning another Feynman insight: biologists already know very well just how densely information can be stored. For example, all the genetic instructions for building your body are stored in the DNA packed into a small part of the nucleus of every one of your cells. And each cell is so small it is invisible to your naked eye.

At this point it may seem to you that we have gone astray. Making extremely tiny photographs is no doubt a good way to save space for library archives and a clever way for spies to pass messages covertly, but what does it have to do with the victory that digital technology has won over the analog alternative? It turns out that a computer these days is essentially a microphotograph.

5.2 The Computer as Microphotograph

It is difficult (but perhaps not impossible) to reduce a computer made with vacuum tubes to microscopic proportions. For one thing, we must think of how we can produce continuous streams of electrons that can move freely in a tiny vacuum. Perhaps "vacuum tubes" can be made small enough so that there aren't enough air molecules to block the possible paths between filaments and plates, but we still must supply the power to get the electrons free in the first place and must also get rid of the heat that would be so generated. It's an interesting exercise to think of how technology might have proceeded without the semiconductor transistor.

As we saw in the previous chapter, semiconductors free us from the mechanical and thermal restrictions of the vacuum tube and make it possible to shrink circuits of valves to microscopic dimensions. This new way to make valves leads, with the development of a great deal of refined technique and complex machinery, to what amounts to the *printing* of computers made of microscopic gates, in much the same way that we print microphotographs. We can break the process down into the following steps, here highly simplified:

- Grow a very pure crystal of semiconductor, usually silicon. Because of the control required of its conducting properties, its purity must be better than one foreign atom per billion semiconductor atoms (99.9999999%, "nine nines").
- Slice the crystal into thin *wafers*.
- Polish each wafer so that it is very flat.
- Coat a wafer, now considered the *substrate*, with some material that we wish to imprint with a layer of a circuit, such as silicon dioxide, say, which is an insulator.
- Coat the silicon dioxide with a special light-sensitive material called *photoresist*.
- Project an image of a predesigned circuit of gates onto the photoresist. This exposes some parts of the photoresist to light and not others, which are masked out.
- The parts of the photoresist that are exposed to light become soluble. Wash those away, leaving the projected pattern as exposed silicon dioxide.
- Etch away the exposed silicon dioxide with a chemical.
- Wash away the remaining photoresist, leaving a pattern of silicon dioxide above the substrate that embodies the projected image as a circuit layer.

This is done layer upon layer, for 20 or 30 layers, and these layers can be interconnected in some stages of the process, making it possible to construct complicated circuits of transistor gates. It is also necessary to dope different parts of exposed semiconductor, which is done at appropriate points in the process by firing high-speed ions at the wafer. The wafer is then sliced into chips and packaged.

Semiconductor fabrication plants, often called foundries, have evolved into today's industrial marvels. The so-called clean rooms where the wafers are processed must be meticulously free of dust and vibration, with carefully controlled temperature and humidity. The high-precision machines that do the microphotography, called *photolithography*—the etching, doping, slicing, packaging, and so on—are expensive, and building a new "fab plant" can easily cost a few billion US dollars. It seems that everywhere we turn in the story of digital computers we run into very large or very small numbers.

5.3 Heisenberg in the Chip Foundry

The complex and precise operation of a semiconductor foundry has, as you might imagine, myriad variations. All that matters to us, however, is that the pattern is imprinted with light— *photons*, which are governed by quantum-mechanical laws. The Heisenberg uncertainty principle therefore sets the rules of the game in the highly competitive industry of chip manufacturing. Because the floor plans of the chips are projected onto the wafers by what is essentially photography, the limit on how small the smallest features can be made is determined by the wave nature of the light that is used for the photography. The size of the smallest detail that can be projected onto a silicon wafer is proportional to the wavelength of the light used: the smaller the wavelength, and hence the higher the frequency, the smaller the detail. Figure 5.1 illustrates this with images of two bright points separated by different distances. The image of each point, called an *Airy disk*, is a widening series of concentric circles caused by diffraction of the light waves, and as the two points are moved closer together (top to bottom) the Airy disks tend to merge, and it becomes increasingly difficult to see that there are two source points and not one.

Because optical resolution is determined by the wavelength of the light used, the historical trend in photolithography has been toward shorter and shorter wavelengths, from deep violet down to deeper and deeper ultraviolets, and a great deal of effort has been put into developing lasers for these applications. There is so much money riding on the ability to produce chips with more transistors on them that no technology is overlooked and no tricks spared in the development of imaging techniques for microchip production. But this optical game has limits that follow from the wave nature of light and Heisenberg's principle— if photons were ideal particles and did not have wave properties, there would be no diffraction of light.

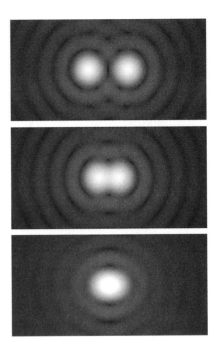

FIGURE 5.1. Diffraction patterns in the image of two point sources, passing through a circular aperture. As the two points get closer together (from top to bottom), they become difficult to distinguish because of the diffraction of light caused by the aperture. (Image by Spencer Bliven via Wikimedia Commons.)

The pressure to pack more and more transistors on a chip has led to the experimental use of electron beams and X-rays in their production, because electrons and X-rays have much smaller wavelengths (the higher their energy, the smaller their wavelength). Thus, the history of silicon lithography may follow that of optical microscopy, with the same fundamental limits.

For example, optical microscopes have no problem showing human red blood cells, which are about 8 μm, or 8000 nm, in diameter.[8] The wavelength of visible light is about 500 nm, much smaller than red blood cells, so blood cells are quite easy to see with an ordinary optical microscope. Even a toy microscope will show red blood cells. However, the influenza virus, a fairly large virus at that, is only about 100 nm in diameter—80 times smaller. The 500 nm wavelength of visible light is therefore much too large to resolve an influenza virus, even with the best optics.

When optical microscopes ran out of steam, electron microscopes came to the rescue—an electron can easily have a wavelength on the order of 1 nm. Beautifully detailed images of the influenza virus are available at a magnification of 100,000, whereas even the best optical microscopes are usually limited to magnifications of about 1500. Once again, it's all a matter of quantum mechanics.

5.4 Moore's Law and the Time of Silicon: ca. 1960–?

In 1965 Gordon Moore, a cofounder of Fairchild Semiconductor and future cofounder of Intel, wrote a short article for a special issue of *Electronics*, a trade journal for the radio-electronics industry.[9] He had been asked to predict the progress of the semiconductor industry over the next 10 years. Despite his many accomplishments, he will no doubt always be known as the eponymous originator of Moore's law.

The data that Moore extrapolated was actually quite meager, a testament to his foresight and intuition. Figure 5.2 shows all that he had to work with. To take a closer look at this data, we list the approximate logarithms (base 2) from his graph:

Year	Logarithm	Number of components
1959	0	1
1962	3	8
1963	4	16
1964	5	32
1965	6	64

The initial jump from 1 to 8 components indicates that it was just after 1959 that it became possible to fit more than one component on an integrated circuit chip. When expressed as a simple list of numbers in this way, it becomes clear what the trend is: the number of components per circuit seems to *double every year*. Moore adopted this doubling rate but with a spot of caution; he says, "Certainly over the short term this rate can be expected to continue, if not to increase. Over the longer term, the rate of increase is a bit more uncertain, although there is no reason to believe it will not remain nearly constant for at least 10 years."

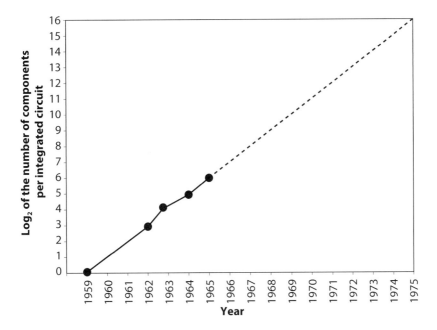

FIGURE 5.2. Moore's very limited data, from his 1965 article. (Courtesy IEEE. Reprinted in *IEEE Solid-State Circuits Soc. Newsletter*, Sept. 2006, 33–35.)

This is how Moore arrived at the figure of 65,000 components in a single integrated circuit in the year 1975, which is his summary prediction.

His assessment of the trend of semiconductor progress in the paper is actually more subtle than just counting the number of components that can be squeezed onto a chip. After all, Moore was a businessman as well as a scientist, and he focused entrepreneurial eyes on the potential of what was then an infant technology.

A key consideration in the production of chips is the *yield*. In a given batch of chips there will inevitably be duds, and the more you push the technology by packing more gates onto a chip, the lower the yield. If you play very conservatively, you get a very high yield but with fewer gates per chip. If you play very aggressively, you get a very low yield but with more gates per chip. There is therefore an optimal trade-off—a sweet spot—that minimizes the manufacturing cost per gate, and it was this measure that Moore used to get the numbers for his projections (see figure 5.3).

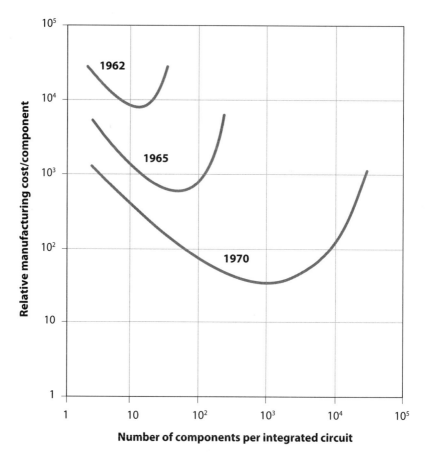

FIGURE 5.3. The trade-off Moore cared about, from his 1965 article. (Courtesy IEEE. Reprinted in *IEEE Solid-State Circuits Soc. Newsletter*, Sept. 2006, 33–35.)

One way or another, all the forms of Moore's law incorporate the history of progress in semiconductor technology and implicitly predict that transistor density will be multiplied by two every one, or one and a half, or two years, or something along those lines. Figure 5.4 shows the actual number of transistors on chips from the early 1970s to the recent past, and figure 5.5 shows the width of the lines etched on those chips to draw transistors, the so-called *minimum feature size*, over the same time span. Considering that Moore wrote in 1965, the accuracy of his prediction seems almost supernatural.

Connoisseurs of chip technology argue nuances. Should we measure the number of gates per chip or the minimum feature

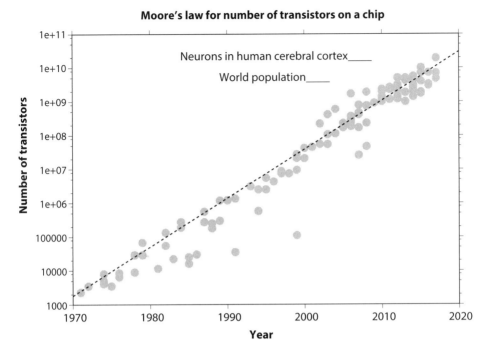

FIGURE 5.4. The number of transistors on commercial chips versus time. The upper limit of this graph is 100 billion, about the total number of neurons in the human brain. (Data from https://en.wikipedia.org/wiki/Transistor_count. Accessed September 11, 2017).

size? How much time does it actually take to change by a factor of two? Is Moore's law really a law in the sense of Newton's law of gravitation? Or is it only a self-fulfilling prophecy propelled by manufacturers' need to meet market expectations? None of this matters to us. What does matter is the doubling in a fixed time, which is the definition of an *exponential* increase in gate density.

There is a classic book by the physicist George Gamow called *One, Two, Three . . . Infinity*,[10] which I first encountered as a boy. Among its charms are the figures drawn by Gamow himself—they were still surprisingly vivid in my mind when I revisited the book recently after half a century. Figure 5.6 shows his sketch of the Grand Vizier Sissa Ben Dahir kneeling before King Shirham of India. Gamow tells us that the king wanted to reward his grand vizier, a skilled mathematician, for having invented the game of

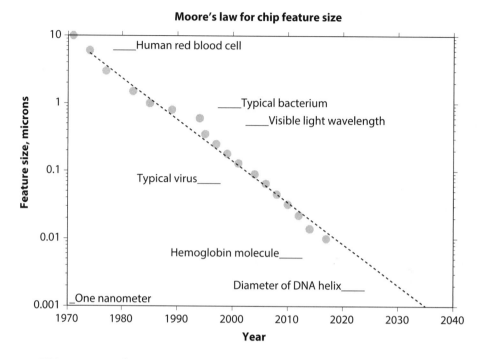

FIGURE 5.5. The minimum feature size used to print a transistor on a silicon chip versus time. Also shown are the sizes of some typical small things. Recall that a micron is a millionth of a meter; a nanometer is a thousandth of that. (Data from https://en.wikipedia.org/wiki/Semiconductor_device_fabrication. Accessed September 10, 2017).

chess. The vizier asked for "a grain of wheat to be put on the first square of the chessboard, and two grains to be put on the second square, and four grains to be put on the third, and eight grains to be put on the fourth. And so on, oh King, doubling the number for each succeeding square, give me enough grains to cover all 64 squares of the board." The king replied, "You do not ask for much, oh my faithful servant, silently enjoying the thought that his liberal proposal of a gift to the inventor of the miraculous game would not cost him much treasure." The request of Sissa Ben Dahir was far from modest. Gamow estimates world wheat production (he wrote in 1947) at 2 billion bushels. Production of wheat is now about 10 times that figure. He assumes that there are about 5 million grains of wheat in a bushel. Using today's 20 billion bushels times 5 million grains per bushel gives us about

FIGURE 5.6. The Grand Vizier Sissa Ben Dahir kneeling before King Shirham, explaining his modest suggestion for his reward for inventing chess; from Gamow (1947). (Courtesy Dover. *One, Two, Three...Infinity: Facts & Speculations of Science.* Viking Press, New York, 1947. Revised 1961, reprinted Dover, 1988.)

10^{17} grains of wheat produced per year in the world. On the other hand, the number of grains on the chess board when the final bushels of wheat would be brought before the king would be about 2^{64}, or about 1.8×10^{19} grains. So the vizier's gift would amount to about 180 years of the world's current wheat production. Such is the power of exponential growth.

Here's another classic example. In a 1969 essay called "The Power of Progression,"[11] Isaac Asimov followed the apparent exponential growth of human population to its logical conclusion. He asked this question: Suppose the human population doubles at the rate of once every 47 years (the rate he quotes for the period 1950–1969). How long would it take for the entire earth to reach the population density of Manhattan? The simple calculation yields 585 years. If that doesn't bother you, Asimov continues, estimating that in 6700 years, doubling at the same rate, all the mass in the entire universe would be converted to human flesh.

Exponential growth in the real world never continues indefinitely. Something's always got to give because the ultimate consequences are always absurd. Kings run out of wheat, people run out of planets, and the United States government stops paying interest on bonds.[12]

5.5 The Exponential Wall

It should be clear by now that I've been setting you up on a collision course with the laws of nature. Heisenberg's uncertainty principle sets a fundamental limit on how fine a line we can draw on a silicon wafer, and quantum tunneling limits how narrow we can make the all-important channel (or equivalent gating structure) in a transistor. Consequently, the exponentially increasing density of semiconductor chips must, sooner or later, hit a brick wall. Consider figure 5.5. If we continue the straight line down, the features we will need to etch to make transistors on a silicon chip will shrink to the distance between silicon atoms in the crystal itself sometime before 2040, and it's a safe bet that transistors on silicon chips won't get to be smaller than that. We are reaching the fundamental physical limits of the current silicon-chip paradigm.

We need have no fear, however, of any "end of progress." Moore's law may die, but the feverish progress in the computer industry must continue. Computers are too useful, people are too dependent on them, and there's too much money to be made in satisfying the demand. This progress can take place on two fronts, *hardware* and *software*, and so far we've concentrated almost exclusively on hardware. The discrete-state idea, with its embodiment in semiconductor crystals, has changed the world, but as we see now, we need new fuel for the fire.

A promising possible direction for hardware progress is offered by quantum mechanics, the same marvelous body of knowledge that prescribes the physical limits of today's conventional, classical computers. But finding a new physical home for computation appears to be a long-term proposition, and so we will leave the hardware front for a while and return to it and quantum computing later.

It is now time to turn to the possibilities, as well as the limits, of software. As usual, our view will be from a high altitude. No code will be written here.

Part II

Sound and Pictures

6 Music from Bits

6.1 The Monster in 1957

Watch someone use a computer today. Notice that she deals with three media: text, sound, and pictures (including video). It is easy to see how text can be stored and manipulated by a computer, because the text is itself discrete in nature. But sound and pictures, although we take them for granted on today's computers, are first of all *analog* in nature. To get back and forth between the analog and digital worlds of audio and video, we need the A-to-D and D-to-A converters mentioned in chapter 1, and these were not fast enough, small enough, or cheap enough until well into the personal-computer revolution.

I have the advantage of some perspective here, and I'd like to set the stage for you with a short flashback to the first computer I ever used, the IBM 704—summer, 1957. Figure 6.1 shows what I believe is the very same machine I used, enthroned majestically in a skyscraper in what was then modestly called "IBM World Headquarters" in uptown Manhattan.

The 704 was one of the last computers to use vacuum tubes instead of transistors, and it filled the space of several large rooms. The tape drives (on the right) were the size of refrigerators, with glass doors the height of the cabinet and vacuum columns at the sides to take up slack in the tape. Running a program was an afternoon's project that started with a cab ride uptown from my office to the machine room shown in the figure, where I punched a time clock to get on and off the machine with my box of prepared punched cards and large spools of tape. When the line printer chugged off my results (I can still smell the ink and see the paper with alternating patterns of horizontal green lines to guide the eye), I collected my decks and output and took a cab back downtown to ponder my results. If there was a bug in the

FIGURE 6.1. The IBM 704, ca. 1954. Your laptop is about 250,000 times faster than this machine and has about 100,000 times as much random-access memory. (Source: *IBM 704 Manual of Operation*, 1954.)

program, it meant another scheduled time slot, another round-trip cab ride, and another afternoon. To run one program.

At the time, the 704 was one of the fastest machines in the world, by definition a "supercomputer." How fast was it? Fortunately, I saved my IBM 704 manual.[1] The basic cycle time of the machine was 12 microseconds, but even the fastest instructions required 2 cycles. Floating-point operations took about 20 cycles, so the basic speed of the computer was about 4 Kflops, where a flop is a floating-point operation per second and a Kflop is 1,000 flops. And just how much faster is the fastest machine today? The fastest supercomputers now use millions of processors in parallel, so to be fair, we'll compare the 704 with my sturdy laptop, which has a clock speed of about 2 GHz (corresponding to a cycle time of 0.5 ns) and might be able to do about a billion flops—the laptop can pipeline the floating-point operations and may even have multiple processors, so it doesn't need 20 machine cycles for floating-point instructions. This makes it about 250,000 times faster than the 704, which means that one second on my laptop would need about three days on the IBM 704 I used in 1957. Not bad for a four-pound device that I can throw in my backpack.

The progress in memory has also been startling. The random-access memory (RAM) for the 704 used magnetic-core memory, which was very expensive and very bulky, and even the largest machines were always strapped for fast memory.[2] The largest size available was 32 Kwords, where each word had 36 bits, which means that the largest of these gigantic machines had a little less than 150 KB of RAM, less than one hundred-thousandth the capacity of a flash drive that today costs less than a bowl of soup in a diner.

These yardsticks of progress are impressive and well known, but the point I'm getting to next is that the 704 not only crawled painfully slowly in very tight quarters, it was also deaf, dumb, blind, and, by today's standards, on the edge of incommunicado.

6.2 A Chance Encounter with a D-to-A Converter

It wasn't until six years later that I saw a D-to-A converter.[3] It happened quite by accident then that two threads came together: I had just started teaching at Princeton after finishing my dissertation on the topic of digital filters, and a couple of very adventurous composers in the music department were trying to coax music out of a bulky D-to-A converter that had just been donated to the university by Bell Laboratories through Max Mathews, a pioneer in computer music. Up to that point, composers of computer music at Princeton had to drive the 80-mile round-trip to Bell Labs to get their digital tapes converted, even more arduous than my round-trip cab drive in Manhattan traffic that I mentioned above.

I happened to pass the room where two composers, Godfrey Winham and Jim Randall, were struggling, and heard the familiar clickety-clack of their tape drive, which was just like the ones pictured in figure 6.1. I also heard some rather alarming sounds, nothing that could be described as music. I poked my nose in the room and asked what they were up to. They told me they were trying to make a digital version of a resonator work, which is something like a tuning fork but with numbers instead of bars of metal. The result, however, was cacophony instead of sonority. I think they were a little more than skeptical when I told them I had spent the last few years of my life thinking about just this sort of problem. After all, I was an unannounced stranger who had just randomly wandered in, apparently off the street, and a rather young and brash stranger at that. But I believe I gained their confidence, and the chance meeting started years of enjoyable collaboration with composers of computer music.

I should mention what was wrong with their digital resonator. The D-to-A converter took as input *fixed-point* numbers, that is, integers. With 12 bits used for each sample of sound, say, the converter expected to receive numbers chosen from 4096 (2^{12}) possibilities. This means that there is a rather restrictive limit to the size of the largest value that the converter can handle, and if it is fed a larger number, it does something quite awful. That was their problem: the digital resonator's output signal went out of the converter's range. The solution to the problem is to *scale* the calculated signal before it is fed to the D-to-A converter, cutting it down to size. Because I had been working on things like digital resonators for my dissertation, it wasn't hard to give Godfrey and Jim the required scaling factor. In fact, it was no harder than a homework problem today in an introductory digital signal processing course. I did have the feeling at the time, however, of having landed on a South Pacific island in a giant silver bird.

6.3 Sampling and Monsieur Fourier

I jumped ahead when I used the word *sample* above, referring to what is fed to a D-to-A converter. If it didn't bother you, it's probably because it is just common sense that if sound is going to get inside a digital computer, in the form of numbers, it will have to be sampled. But what, exactly, is sampled?

Sound is transmitted in the air through alternating waves of compression and rarefaction, which are called *longitudinal* waves. This is in contrast with *transverse* waves, where the local motion is from side to side, as, for example, on a guitar string. You can excite both kinds of waves on a Slinky, depending on whether you push a free end or wiggle it laterally. A microphone converts the pressure wave in the air to a voltage signal, and it is this signal that is sampled—almost always at regular time intervals.

We are then faced with the natural question, How fast do we need to sample a sound wave (as represented by the voltage signal from a microphone) to represent the sound faithfully? How many samples per second? There is a very beautiful and easily stated criterion for this, based on frequency content, stemming from the work of Jean-Baptiste Joseph Fourier, almost two hundred years ago.

We can regard any signal, including of course a sound signal, as being composed of a sum of different frequencies. This is a profound idea, which we encountered in our earlier discussion of noise, and it is worth taking a moment to sketch some of its ramifications. Without going into the mathematics, it turns out that we can view any signal, like the voltage signal from a microphone, as either a plot versus *time* or versus *frequency*. The more technical terminology is that we can view the signal in either the *time domain* or the *frequency domain*. The name of the rule that gets us from the time domain to the frequency domain is called the *Fourier transform*, and the reverse rule is called the *inverse Fourier transform*. In this way we can move freely back and forth between the time and frequency domains with no loss of information.

It is also important that certain operations on a signal in the time domain correspond to other operations in the frequency domain. Loosely speaking, everything that happens in the time domain also happens—through an appropriate kind of lens—in the frequency domain. In such situations we say that there is an *isomorphism* between the two domains.[4] The lens analogy is not fantastical. An ordinary glass lens can, in fact, be used to find Fourier transforms of images regarded as spatial (two-dimensional) signals.[5]

6.4 Nyquist's Sampling Principle

A signal in the real world is always limited in how high a frequency it can contain. The reason is similar to the reason that transistors are limited in their speed of operation; all electronic devices have a certain amount of capacitance, which limits the speed with which charge can accumulate, which limits the speed with which voltages can change. Mechanical devices have a corresponding amount of inertia. These factors limit the highest frequency that a real signal in any particular physical environment can contain.[6] The point is that we need to worry only about sampling the *highest* frequency in a signal. The lower frequencies are easier, not harder, to represent at a given sampling rate.

We're now ready to derive, admittedly in a heuristic way, the criterion mentioned above for adequately fast sampling. The idea of a "pure" tone of a given frequency is often introduced as a *sine wave*, the familiar waveform that goes up, levels off, then goes down, levels off, and so on. The function *sine* is referred to as a "circular" function, for the following reason: Picture a rotating circular disk, horizontal, a roulette wheel if you like, with a point of light (from an LED, say) glued to a fixed point near its edge. If we darken the room, we see the light rotating continuously at a given rate, at a certain frequency in "cycles per second," or Hz. If you kneel down and look at the disk from the side, the light goes back and forth, and, in fact, it will describe precisely the waveform called a sine wave. This is a great convenience, because we can now think about the rotating disk, which is much easier to visualize, and a more precise picture than an undulating wave. As an aside, I point out that physicists and engineers make heavy use of this alternative representation of a sine wave, albeit mathematically, in the form of a complex-valued function called a *phasor*. Richard Feynman wrote a marvelous little book, called *QED*, explaining quantum electrodynamics in simple terms, and he uses the picture of a little spinning disk throughout.[7]

Now, instead of leaving the LED on steadily as the disk rotates, flash it periodically. Each flash corresponds to a sample of the position of the little light as it turns with the disk. If we sample many times for each rotation of the disk, we have no trouble representing the true rate at which the disk is turning. However, if we try to get away with slower sampling, we reach a point where we are sampling *exactly twice* for each rotation of the disk, and the little point of light will just flip back and forth between two positions 180° apart. If we now try to get away with slower sampling, flashing (sampling) the light a little less often than twice per rotation, slower and slower, a rather bad (but interesting) thing happens: the little light appears to turn in the direction *opposite* to its actual direction. If we slow the flashing down to only once per rotation, the flashing light appears stationary. If we flash even slower than once per rotation, the little light appears to start rotating in the correct direction but at a very slow rate—much slower than the true rate of the disk.

This is exactly what happens in an old Western movie when the stagecoach draws to a stop. The wagon wheels appear to turn in the wrong direction, slow down, start turning in the right direction, and so on, until they appear to be turning in the correct direction, more and more slowly, until they finally draw to a stop. The sampling behind this phenomenon is the frame rate of the movie camera, which is standardized at 24 frames per second. When the wagon wheel is turning faster than 12 times a second, we are in effect sampling at a rate less than twice per rotation, and the image shows a fraudulent representation of the speed of the wheel. In fact, practitioners of digital signal processing (DSP) call such a fraudulent frequency an *alias* of the true frequency.[8]

We can now draw the promised elegant conclusion from this imaginary experiment: To capture faithfully the frequencies in a signal, we must sample at a rate at least twice the highest frequency present in the signal. Put the other way around, if we sample at a given rate, we must limit the highest frequency present in the signal to *half* the sampling rate. This latter rate is now called the *Nyquist frequency*.

Harry Nyquist worked for Bell Telephone Laboratories, which was very concerned with communication problems from the early days of the twentieth century, for obvious reasons.[9] He explains his principle in Nyquist (1928a), but his explanation is in terms of telegraph terminology that is 90 years old, and he is not always easy to interpret. But Nyquist's principle, sometimes called his sampling theorem, is there.

What this means in our modern world, for example, is that audio signals, which are usually limited to frequencies (well) below 20 kHz, need to be sampled at a rate of at least 40 kHz.

In fact, the standardized sampling rate used by compact discs is 44.1 kHz. Exactly the same ideas apply to A-to-D conversion of video signals, but the rates are much higher.

6.5 Another Win for Digital

Notice that Nyquist's principle is stated as a *necessary* condition. It tells us that we must sample at a rate at least twice the highest frequency in a signal, but it doesn't guarantee that sampling at this rate will result in a particularly accurate representation of the original analog signal. It is an amazing fact that sampling at twice the highest frequency is not only necessary but sufficient to determine the original signal *perfectly*.[10]

This much more powerful and consequential version of Nyquist's principle was stated 21 years later in precise and general terms by Claude Shannon—another enormously influential Bell Laboratories researcher, and one whom we shall meet again very soon. In Shannon (1949) he states it as "Theorem 1: If a function contains no frequencies higher than W cps [Hz], it is completely determined by giving its ordinates at a series of points spaced $1/(2W)$ seconds apart." In fact, the result is sometimes called the Nyquist-Shannon sampling theorem.[11]

I very much like Shannon's heuristic observation, which is related to the flashlight-on-a-disk argument above but is not exactly the same. He states the principle in terms of the *period* $1/W$ of the highest frequency W in a given analog signal: "This is a fact which is common knowledge in the communication art. The intuitive justification is that, if [the analog signal] contains no frequencies higher than W, it cannot change to a substantially new value in a time less than one-half cycle of the highest frequency." That is, at a rate $2W$, twice the highest frequency.

As the section title says, this is a big win for the digital way of doing things. It tells us that if we sample at the rate required by Nyquist's principle (or faster), we can in principle reconstruct the original analog signal *perfectly* from its samples. In theory we can do anything in the digital domain that we might want to do in the analog. It's worth pondering the point a bit; it justifies much of what we today call DSP.

There are, of course, unavoidable imperfections in this process, and I'll just mention a couple of them, briefly. They are not a critical part of the picture because their effects can be made as small as we wish, it turns out, by using more speed and storage.

The first imperfection is the necessarily limited accuracy in measuring the values of the samples we take of an analog signal. The standard adopted for compact discs is 16 bits, which means there are 2^{16}, or 65,536, possible different levels that can be distinguished. Higher accuracy is possible but ordinarily not worth the trouble and expense. Noise is always lurking somewhere in your audio system—in the microphone and preamplifier electronics, in the acoustic environment, the background, or anywhere else in the analog part of the system—and if you go much beyond 16-bit resolution, you are wasting a lot of effort capturing noise that is usually inaudible.

Another imperfection in the A-to-D process, one that became apparent as a potential problem early in the history of digital audio, stems from the requirements of Nyquist's principle. As we described in the flashlight-on-disk picture, frequencies that might be present above half the sampling rate—the Nyquist frequency—will be "aliased down" below the Nyquist frequency by the sampling process, and sounds at those unwanted frequencies can be quite troublesome. Quite awful for music, in fact, because in general the frequencies in those unwanted sounds bear no harmonic relation to the pitches in the original analog sound. In practice, therefore, an analog signal is filtered before it is sampled to eliminate frequencies above the Nyquist frequency, a process called *lowpass prefiltering*. But it is a fact that no such filtering can do a perfect job blocking the unwanted high frequencies. We just need to filter well enough before sampling to reduce aliasing to an acceptably low level.

By the way, I have focused on sampling audio, although I reminded you from time to time that sampling video works the same way. To be honest, one reason is that audio processing is just closer to my heart. The main reason, though, is that digital imaging is more complicated because it requires sampling in two dimensions. In fact, digital television entails sampling in three dimensions (the image is moving), and the development of digital television lagged behind digital radio by about 20 years, even with the explosive growth of hardware that resulted from the progress of Moore's law.

Digital imaging provides a familiar example of aliasing, which in imaging is called the *moiré effect*. For example, a shirt with narrows stripes will, under the right circumstance, shimmer with wavy patterns, because the stripes are at frequencies above the Nyquist frequency, and so will be aliased to lower frequencies. When the shirt moves with respect to the camera, the angle changes and the aliasing shifts continuously. You can also run into aliasing, for the same reason, when you scan documents that are formed of small dots, like newspaper photos. Scanners usually have software to ameliorate the effect; the software filters the original photo to suppress high frequencies, in exact analogy to the lowpass prefiltering used in audio systems. Digital cameras can accomplish the lowpass filtering by effectively blurring the image, slightly of course, in one way or another.

Some of the interesting and useful aspects of DSP have to do with clever ways of dealing with accuracy and aliasing problems. For example, it is often possible to trade off more speed for less accuracy, and it is sometimes more cost-effective to make devices faster and less accurate.

6.6 Another Isomorphism

To take stock, we see that we can operate on signals, video as well as audio, just as well in digital as in analog form. In fact, just as we can view signals in either the time or frequency domain, we can view signals in either the analog or digital domain. We've seen one example of this when we considered sampling at twice the highest frequency in a signal. In this case, by Nyquist's principle, we can go back and forth between an analog signal and its sampled version with (in principle) no error at all. The upshot of Nyquist's principle is that there is an isomorphism between the domains of analog signals with limited frequencies and the sampled versions of them.

By the way, this is not the only isomorphism between analog and digital signals. There is one that does not require the analog signal to be limited in the frequencies it contains.[12] You might wonder how the isomorphism avoids aliasing ... well, it does not use sampling but rather some other, more complicated way to get a digital signal from an analog one. The details aren't important here. The point is that analog and digital signal processing are equivalent in quite a general sense, which is really why we can take sound and pictures for granted on the digital devices that surround us today.

7 Communication in a Noisy World

7.1 Claude Shannon's 1948 Paper

Go back and watch the computer user of today, the one we spied on at the beginning of the previous chapter, the one who uses text, sound, and pictures. Chances are it won't be long before she sends or receives information to or from some other computer, quite possibly hundreds or thousands of miles away. Such communication requires fiber optic or copper cables, or radio, all of which put a definite limit on the rate at which information can be sent from one place to another. In fact, there is always a limit on how fast information can be sent via *any* medium. Why is this so? The answer returns us to a recurring theme, and to earlier discussions of the limits of analog computation and microchip fabrication: the world is, essentially and unavoidably, a noisy place.

Naturally, the general problems of communication through a noisy medium attracted the attention of the phone company's formidable research facility, Bell Laboratories, which, you will remember, employed both Nyquist and Shannon. Two decades after Nyquist described his sampling principle, Shannon did something quite rare in science generally: he single-handedly and at one stroke founded a brand-new field, known as *information theory*. As the Soviet mathematician Aleksandr Khinchin remarks, "Rarely does it happen in mathematics that a new discipline achieves the character of a mature and developed scientific theory in the first investigation devoted to it."[1] The remarkable publication he is referring to is Shannon (1948).

Shannon's 1948 paper, actually in two substantial parts, did exactly what Khinchin said: it established a full-fledged field in one brilliant blow. The field is also a bit peculiar because it is at the same time both a part of mathematics (more specifically probability theory) and communication engineering. To see how information theory provides yet another reason why the digital way of doing things has replaced the analog, we review, in our customary informal and nonmathematical way, the central and, actually, quite astonishing result of Shannon, often called his *noisy coding theorem*. It's about the rate at which information can be transmitted through a channel that is noisy, and so we must first describe how "information" is measured.

Before going on, however, I need to draw an important distinction. In speaking of a fundamental limit on how fast information can be sent from one point to another, we must distinguish between the *rate* at which information can flow, as water in a pipe, and the *delay*, or *latency*, between the sending and receiving of a particular bit. In the latter case, we know the fundamental limit is the speed of light. In the former case, which is usually the limiting factor for consumers of streamed data, it is Shannon's noisy coding theorem that is in play. When you test the speed of your internet connection, for example, you are usually worried about the rate, measured in millions of bits per second, rather than the latency, measured in thousandths of a second.

7.2 Measuring Information

To see how we might go about measuring information, let's think about flipping a "fair coin," which means a coin that is not biased toward either heads or tails. In common parlance, we say the chances of heads or tails are even, or "fifty-fifty." More formally, we say the *probability* that heads or tails comes up is one-half. Weather forecasters make good use of the statistical hedge: a casual look at a meteorological site tells me that "A strong and fast-moving tropical system over the east-central Caribbean Sea has a 70% chance of developing into a tropical cyclone."

If I flip a coin once, and I don't tell you the result, I introduce a certain amount of uncertainty in your head. If I then tell you the result, I remove that uncertainty. We say I have given you some information. It is a simple, but important, insight that *information is the removal of uncertainty.* How much information have I given you? In the simple case of flipping a fair coin, this is easy: We say that the information in the result of flipping a fair coin is one *bit*.

If we flip the coin twice, we say the information about the result is two bits—provided that the second flip is not affected in any way by the first flip. Three independent flips, three bits, and so on.

A useful way to think about a sequence of events like coin flips is to consider the total number of possible outcomes. In the case of one flip, there are two equally likely outcomes. In the case of two flips, four equally likely outcomes. In the case of three flips, eight equally outcomes, and so on. You can see what is happening here. The amount of information is the number of times I need to multiply one by two to get the number of possible outcomes. Each flip multiplies the number of possible outcomes by two. There is another name for "the number of times I need to multiply one by two to get a number," and that is the logarithm, or log (base 2), of that number. We stick to logarithms base 2 here, but we could just as well use logarithms base 10, say. Changing the base, however, introduces a scale factor and changes the unit of information. For example, using logarithms base 10 yields information in units of *decimal digits* (as you might guess) instead of bits; one decimal digit is worth about 3.322 bits.

Now suppose something happens somewhere in the world—the flip of a coin, a tropical cyclone—and suppose you are unaware of it. When I tell you the result by sending you a message, we say that I have given you an amount of information equal to the number of times we need to multiply one by two to get the total possible number of equally likely messages that you could have received. You can therefore think of the amount of information in a message as the *equivalent number of coin flips* in the receipt of the message. If I flip a *fair* coin 10 times, there are 2^{10} possible messages that I could send to you: heads or tails on the first flip, heads or tails on the second, and so on. The information content of such a message is the log of 2^{10}, or 10 bits.

So far, we have discussed how to measure information when we are dealing with equally likely events. What about that cyclone, with a chance of 70% (or, what is the same thing, a probability of 0.7)? How much information do I send you when, two days from now, I tell you that a cyclone did or did not develop? We next derive a measure of information for this case, when we are dealing with two events that are not equally likely. The measure turns out to be very natural—and unique, in the sense that no other measure (up to a scale factor) has the properties that we want. Standard textbooks on information theory usually define the measure and then prove its properties,[2] but we will instead be satisfied to motivate it in an informal way.[3]

The key idea is to think of the 70% chance of a cyclone as being derived from 100 *equally likely* (hypothetical) possibilities, 70 of them positive and 30 of them negative. Any probability can be broken down this way. To take another example, if the probability of an event is one-third, we can think of three equally likely events, one counting as positive and two as negative.

Now, to continue with the example of the possible cyclone, suppose two days from now a cyclone does develop. You are not interested in exactly which of the 100 hypothetical events occurred, but only in the fact that one occurred among the 70 "positive" ones. If I send you a message telling you exactly which of the 100 events occurred, I send you log 100 bits, but that is much more than necessary. In the event that there is a cyclone, I have sent you extra information in the amount log 70, giving you the irrelevant information pinpointing which of the 70 "positive" events occurred. So the information in the message "a cyclone occurred" is log 100–log 70. It may be a while since you were introduced to the wonders of logarithms, so I will take the liberty of reminding you that subtracting logarithms divides; the information in the message that the cyclone occurred is $\log(100/70)$, about 0.515 bit. Thus, if an event has a 70% chance of happening, the news carries with it only about half a bit of information.

If you look back on what we just did, you can see that in general the information in a message informing you of an event that has probability p is $\log 1/p$. To check this against our earlier discussion, when I send you a message telling you that a single flip of a fair coin resulted in heads, I am sending you $\log(1/0.5) = \log 2 = 1$ bit.

We can also check that this at least conforms to intuition in extreme situations. If an event is very likely, its probability will be near one, which means that the message announcing it bears very little information. For example, the probability of the sun rising tomorrow is very close to one, say (not too optimistically, I hope), one minus one-trillionth.[4] The information that the sun does indeed rise turns out to be 1.44 trillionth of a bit.[5] Not exactly headline news. On the other hand, consider the event that the sun does *not* rise. That has probability, by our estimate, of one-trillionth, and the information content in a message that this disaster has occurred (a headline indeed) is the log of a trillion, or about 40 bits. This may not seem like much, but consider how much more difficult it is to predict correctly the result of flipping a fair coin 40 times in a row, compared with predicting that the sun will rise tomorrow.

7.3 Entropy

Going back to the example of an uncertain cyclone, it is easy to find the *average information content* in a forecast, as opposed to the information content of a particular message. We've calculated that the information content in a forecast stating that a cyclone has formed is 0.515 bit. That is an event with probability 70%. A negative message has probability 30%, which corresponds to 1.74 bits. The average information in the kind of weather forecast we are discussing is then 0.515 bit 70% of the time, and 1.74 bits 30% of the time. This weighted average turns out to be 0.881 bit. We call this the *entropy*, or *self-information*, of the weather report when it is forecasting cyclones—given that the long-term chances of a cyclone are 70%.[6]

The entropy of a message that tells you if the sun has risen is 40 bits times the probability that the sun does not rise (one-trillionth), plus 1.44 trillionths of a bit times the probability that the sun does rise (one minus one-trillionth), so the entropy is about 41.4 trillionths of a bit. Not an information source you are likely to pay much for.

Here's one more example, with numbers rigged to be easy, from Cover and Thomas (1991): Suppose a horse race has eight competing horses, and the probabilities of winning for the eight horses are 1/2, 1/4, 1/8, 1/16, 1/64, 1/64, 1/64, 1/64. The entropy is $(1/2) \times \log(2) + (1/4) \times \log(4) + \ldots$, which turns out to be 2 bits. The news that a particular 1-in-64 long shot has won is a high-information message (6 bits); on average, though, news of the winner of a race carries 2 bits, the entropy of the horse race.

Before going on, I want to take note of the fact that we have slipped, without much fuss, into talking about probabilities and random events. This is inevitable if we define information as the removal of uncertainty, since randomness is just a way of characterizing uncertainty. Before we toss a die, we are uncertain about which of the six faces will come up, and we say that the result of tossing the die is a random event. It was the same story when we described the (uncertain) noise that corrupts analog signals. In fact, the presence of noise in transmission systems of all kinds has made probability theory a basic tool of communication theory, and it was one of Shannon's most important contributions to recognize that information is basically statistical in nature.

A note on dreamy-eyed science

It seems that every few years a scientific term catches the fancy of readers of scientific bent, and *entropy* is a venerable example. The *chaos* and *black hole* vogues are more recent examples. With some justification: Julia sets and fractals are intriguing, and black holes make for exciting adventure stories in outer space.

When Shannon chose the term *entropy* for his measure of information, the word itself and its mathematical form had already been in use in scientific circles since the mid-nineteenth century, and the Austrian physicist Ludwig Boltzmann had used it to formulate the second law of thermodynamics in the 1870s.[7] When as a child I made a friend of Gamow's book,[8] which you will remember from the Grand Vizier Sissa Ben Dahir's modest request for grain, there was much talk of entropy increasing until the "heat-death of the universe," and the second law was held responsible for defining the "arrow of time." In fact, Gamow's book has a section called "The Mysterious Entropy," reflecting, I suppose, the fashion of popular science of the time. His discussion, however, is demystifying in the best tradition of masterful scientific storytelling.

Fortunately, we don't need to delve into the meaning of entropy in thermodynamics and its relationship to Shannon's measure of information; we can proceed happily with its definition as the average information in a message—the average value of the logarithm of $1/p$.[9]

7.4 Noisy Channels

Shannon's development of information theory is sleek, proceeding in a straight line, with just a few basic concepts. We've defined the information content of a source of random messages. These messages are transmitted over a *channel*, which, because it is in general corrupted by some kind of noise, is not perfectly reliable. The *output* of the channel is therefore another random variable, which is, we hope, closely related to the original input. There are two sources of randomness here: first, the original message, which is an information source that we consider the *signal*; and second, the noise that can cause errors in its transmission over the channel.

The basic question is, Given a maximum allowed rate of transmission errors, presumably quite low, how much information can we pass through a given channel per unit time? The answer has important consequences. It determines whether it's practical to carry on a videoconference over a particular internet connection, or how long it takes to download a photo of Saturn from a satellite orbiting the planet. Coming close to answering such questions requires a university course or two, but our job here is only to contrast the analog and digital ways of doing things.

Channels in the real world, dealing as they do with radio waves and electrical or light pulses on cables, are analog by nature, just as sound that we hear with our ears and images we see with our eyes are inherently analog. But signals are usually converted from digital to analog at the sending end of a channel and converted back from analog to digital at the receiving ends, for reasons that we are about to expand upon.

To take a concrete example, suppose that you send me a voicemail message from your smartphone. What happens to it along the way? First, the sound of your voice is converted from an analog pressure wave at a tiny microphone to digital form and stored in your phone as a sequence of bits, as we have already discussed. At that point it is processed in any number of ways, perhaps filtered to emphasize or de-emphasize different frequency bands, or compressed in some way, without distorting it too much, so that it is shorter. The bits that represent your voice are then packaged in chunks (packets) and used to modulate the analog radio signal that is transmitted from your phone to a cell tower. At the cell tower, the analog radio wave is again converted to digital form and processed in various ways—perhaps filtered or cleaned up in some way, interleaved with other signals from other phones, or perhaps stored while waiting for an available outgoing slot. Once again, the current form of your voice signal is used to modulate an analog radio wave that leaves the cell tower, or perhaps an analog electrical or optical signal that leaves the tower through an underground copper or fiber cable. And so it goes, with your voice signal finally ending up as bits stored on my smartphone, waiting for me to bring it up as a voicemail message and listen to it after a final digital-to-analog conversion.

Every time something in the least way complicated needs to be done to a signal, it is converted from analog to digital form, processed, and then converted back if it needs to be transmitted by radio or cable, or listened to (or watched, in the case of a video signal). This bouncing back and forth between analog and digital form can happen scores of times between your voice and my ear. The basic reason for all these conversions to digital form is that digital processing is cheap, flexible, easily programmable, and, as we've seen from the start, essentially error-free because of its discrete nature and signal standardization.

The analog links in the chain, however, are relatively error-prone, and understanding the consequences of noise is where information theory shines.

7.5 Coding

Sending a bit in the form of a radio wave or a pulse on a cable is a risky proposition compared with processing it on a computer. It is worth repeating that digital processing, because of the standardization of discrete states, is essentially error-free, at least compared with radio and cable transmission over distances

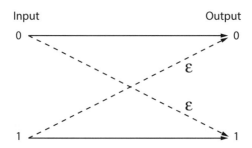

FIGURE 7.1. A binary symmetric channel. The input **0** or **1** is on the left, the output is on the right, and an error in transmission (dashed arrow) occurs with probability ϵ.

vastly greater than the size of semiconductor chips. Errors in transmission, on the other hand, are not nearly as rare as processing errors, especially when you are in a low-signal or high-noise area, like a tunnel or basement. For example, suppose your cell phone transmits a **0** using its radio transmitter. The probability of the cell tower receiving a **1** in error is much, much greater than the probability that a **0** inside the digital part of your cell phone somehow gets mistaken for a **1**. The latter can happen, of course, but only because of some exceptionally rare noise pulse or glitch in the electronics. This superiority of digital processing is traceable to the same root cause that we observed in chapter 2: the digital state is constantly being restored to one of two discrete values, whereas the analog state is continuously exposed to corruption by noise.

The standard way to deal conceptually with a real (imperfect) analog channel is to posit that **0**s and **1**s are being transmitted, and that there is a certain probability, traditionally called ϵ (Greek "epsilon," the time-honored symbol for a small quantity), that a **0** gets inadvertently changed to a **1** and vice versa. This model is called a *binary symmetric channel* and is sketched in figure 7.1. The term *symmetric* refers to the fact that we assume (only for simplicity) that the probability of an error from **0** to **1** is the same as the probability of an error in the reverse direction. This highly idealized model captures the essence of a noisy channel surprisingly well; it is invariably used in the beginning study of information theory and certainly suits our purpose here.

Keep in mind that transmission errors are caused by the inevitable noise on the analog channel. It was realized early on that introducing redundancy in the transmitted signal, in one way or another, makes it possible to detect and correct those errors. Any scheme for incorporating redundancy in the transmitted sequence of bits is generally called a *code*, and, as you might expect from its importance in our lives, coding theory has developed into a highly sophisticated science.[10] It would not be too much of a simplification to say that information theory has two main parts: the first knowing what is possible, and the second getting close to it by coding. I'll give two elementary examples of codes, to get the flavor.

A single-error detecting code

Some codes are designed to allow just the *detection* of errors, with no thought of correcting them. We can accomplish this with the simple and well-known device of a *parity bit*. Suppose we are sending blocks of three bits. We can add a fourth bit to each block that makes the total number of ones even (say). If we then receive a block of four bits with an odd number of ones, we know that an error in transmission must have occurred. We would under this circumstance have no idea which of the bits is in error or, in fact, whether or not three errors might have occurred. The best we can do is discard the block, and, if we can, ask for a retransmission.

This scheme is effective for small blocks, but as the block length gets long, it becomes more and more likely that an even number of errors will occur, and such events will escape detection. The size of the channel error probability, ϵ, therefore limits how long we can make the blocks. But shorter blocks mean that we are sending a larger fraction of bits to check parity, and this results in a slower overall rate of transmission. For instance, in the example just given, we need to send a total of four bits for every three original signal bits, so 75% of the traffic is used for actual signal. In contrast, if we add a tenth bit as a parity check to blocks of nine signal bits, 90% of the traffic is used for signal. But we would in the latter case be more susceptible to double errors

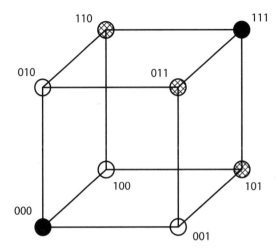

FIGURE 7.2. Looking at a single-error correcting code as the labeling of a cube. The labeling is such that adjacent vertices differ in one bit. The first bit of the code tells us whether we are on the front or back face; the second bit whether we are on the bottom or top face; and the third bit whether we are on the left or right face. The open circles indicate the possible received messages when **000** is sent with a single error, while the hatched circles indicate the possible received messages when **111** is sent with a single error. Assuming that only single errors can occur, if we receive a message corresponding to an open circle, we know that **000** was sent. And, similarly, if we receive a signal corresponding to a hatched circle, we know that **111** was sent. We thus can correct single errors.

escaping detection. As we shall see shortly, this trade-off between the true transmission rate and the error rate seemed, before 1948, fundamental and inescapable. That's why Shannon (1948) was such a big surprise.

A single-error correcting code

It is also possible to cook up codes that enable the receiver to *correct* errors as well as detect them. Figure 7.2 shows the simplest example in geometric form. Suppose we want to send just a single bit, a **0** or a **1**. Think of these as the two opposite vertices of the cube in the figure, where they are indicated by solid dots, labeled **000** and **111**, respectively. Using the code means that when we want to send the message **0**, we actually transmit **000**, and when we want to send the message **1**, we transmit **111**. Assume in the following discussion that only single errors can occur. If two errors can occur, all bets are off.

Now notice that because of the way we labeled the vertices in figure 7.2, a single error when we transmit **000** (for a **0**) will move the received message from the vertex labeled **000** to one of the three vertices labeled **100**, **010**, or **001**. Similarly, if we transmit **111** (for a **1**), a single error will move us to one of the three vertices labeled **011**, **101**, or **110**. Therefore, if we receive a block with a single one, we know that **000** was transmitted and the message **0** was intended. If we receive a block with two ones, we know that **111** was transmitted and the message **1** was intended. As promised, the code enables us to detect *and* correct single errors, but at the price of having to transmit three bits for every message bit.

7.6 The Noisy Coding Theorem

And so, as I've suggested, communication engineers in the years immediately following World War II were in the dark about what was possible using noisy channels. (Blahut 1987, p. 6) put it this way: "Before Shannon's [1948] paper it was generally believed that noise limited the flow of information through a channel in the sense that ... as one decreased the required probability of error in the received message, the necessary redundancy in the transmitted message increased—and hence the true rate of data transmission decreased." The picture drawn was that, as we code to decrease the probability of error, the transmission rate tails off to zero. This is simply false, and how Shannon got at the truth is one of the remarkable intellectual leaps that we encounter repeatedly in this book.

The actual state of affairs is summed up in what is called the *noisy coding theorem* .[11] The theorem goes, informally, as follows: A noisy channel has associated with it a certain *capacity*, C, in bits per second. It is possible to transmit (using appropriate coding) at any given rate below C with arbitrarily small error rate. Conversely, we can reduce error in this way only at rates below C.

The noisy coding theorem explains the indispensable role that information theory plays in giving guidance to the communication systems designer. It tells her that, over a particular channel, she can aspire to transmit with good error performance up to the capacity rate but no faster. Having a theoretical benchmark like this helps the designer of communication systems in the same way that the laws of thermodynamics help the designer of power plants or, as we shall see later, complexity theory helps the designer of algorithms. Knowing what is possible and what is not possible is enormously useful.

Those who live on Earth know that no great boon comes without a price, and the price we pay for transmitting with a vanishingly small rate of errors is in the coding. The proofs of the noisy coding theorem require that we code in *blocks*,[12] taking bigger and bigger gulps of the input data as we approach the promised low error rate. The drawback to this is that it introduces delay in the transmission, which may or may not be a serious problem, depending on the particular situation. For bulk data, delay may not matter. But for telephone conversations, for example, there is a definite limit on how much delay customers will tolerate. There are also inevitable trade-offs between the complexity and effectiveness of codes, and these present problems of design that continue to provide employment for communication engineers.

As mentioned, information theory was born in an exceptionally coherent form. From its central result, the noisy coding theorem, we take away the concept of channel capacity and the potential power of coding to achieve arbitrarily small error rates—but only when transmitting below that capacity.

7.7 Another Win for Digital

As you might expect, there is also an analog version of the noisy coding theorem, and the appropriate analog capacity of a channel can be achieved by analog coding. The reason why, once again, digital processing beats analog processing is anticipated by the following remark of Robert Gallager: "In recent years, the cost of digital logic has been steadily decreasing whereas no such revolution has occurred with analog hardware... This is not to say, of course, that completely analog communication systems are outmoded, but simply that there are many advantages to a primarily digital system that did not exist ten years ago."[13] That was in 1968. A half century of Moore's law has made this observation, albeit in hindsight, sound almost sarcastic. Today any coding or other signal processing at the sending or receiving end of transmission is, except for the crudest kinds of essentially free pre- and postfiltering, digital.

By the way, you may recognize an echo of information theory in the now popularized term *bandwidth*, a term that neatly embodies the idea that the ability to communicate at a given speed is somehow a fundamental and essentially costly commodity. Its use is justified by the fact that the capacity of a channel is, under appropriate assumptions about the noise, directly proportional to the width of the band of frequencies used.

Part III

Computation

8 Analog Computers

8.1 From the Ancient Greeks

We have seen why the very fundamental idea of handling information in discrete rather than continuous form has quite suddenly made information processing our dominant technology. Restricting ourselves to two states means that signals can be made practically immune to the effects of the noise that permeates the analog (some would say real) world. It also enables us to shrink electronic circuitry to spectacularly small sizes. And, as we have just seen, it makes possible essentially perfect exchange and storage of sound and images over channels that are themselves very far from perfect. With this we have set the stage . . . but it is *computation* that is the soul of the revolution, and it is time to discuss computers!

Before computers weighed just a few ounces and became indispensable accoutrements of the well-appointed human, they were regarded only as problem-solving tools. The earliest were, of course, analog machines. Their principles of operation are very diverse, and many of them are quite ingenious, designed to solve very particular, pressing problems. They work by using some physical system—mechanical, electrical, hydraulic, optical—that follows the same rules of behavior as whatever it is we are interested in studying. Hence the term *analog*. Because there are so many interesting problems to solve, and so many interesting things to fool with, the story of analog computers is intertwined with the history of all of science and mathematics. Next, we sample a few high points in this story.

As wide-ranging as the possibilities are for devising analog computers, I impose, for now, one important restriction. We restrict our attention to devices that rely for their operation only on *classical* physics; that is, physics before quantum mechanics, which means before the twentieth century. We return to the important subject of quantum-mechanical computers in chapter 11.

For a very simple example of an analog computer, we go back to the fifth century BC and the astronomer Meton of Athens, who observed that 19 years is very close to 235 (lunar) months—within a few hours, actually. This is a very convenient fact for those who want to relate solar and lunar calendars, and the 19-year cycle, called the *Metonic cycle*, has been exploited by many cultures. Now suppose we make two gears that mesh, one with 19 teeth, and another with 235 teeth. If the gears are engaged and one is cranked, the shaft of the 19-tooth gear will turn 235 times for every 19 turns of the 235-tooth gear.[1] Thus, we can consider revolutions of the 19-tooth gear as counting months (Moon around the Earth) and revolutions of the 235-tooth gear as counting years (Earth around the Sun). With this we have built an analog computer that mirrors the motions of the Sun and Moon, and in this way displays their positions in orbit and their relative phases.

To market our lunar-solar machine, we might want to put it in a stylish wooden box and connect the gears to pretty dials for display and . . . well, that brings us to the first stop on our little tour.

The Antikythera mechanism

The first computer worthy of the name is known only from a single example that went down in a Roman shipwreck sometime around 70 BC near the Greek island of Antikythera, which sits in the Aegean Sea between Crete and the Peloponnese. The wrecked ship was discovered in 1900 by a group of sponge divers, and pieces of the *Antikythera mechanism*, as it is now known, were not noticed until nearly eight months after the subsequent excavations were completed.[2] Unfortunately, two thousand years of salty Mediterranean water had transformed the intricate mechanism into an assortment of corroded, encrusted fragments, and scholars have been at the painstaking work of reconstructing the machine ever since. A recent account of progress in understanding the mechanism is given by Freeth et al. (2006), who remark that it is "technically more complex than any known device for at least a millennium afterwards."

While gaps remain in the current best reconstruction of the Antikythera mechanism, what we do know about it is remarkable. The device consists of a clockwork of at least 30 intermeshing gears with different numbers of teeth, presumably hand-cranked, connected to dials, one on the front and two on the back. As the crank is turned, the indicator dials show the movement of the sun, moon, and probably the five then-known planets, through the zodiac, as well as the occurrence of lunar and solar eclipses. As we might expect, the Metonic cycle used as an example above plays a central role in the operation of the mechanism.

The Antikythera mechanism was far from a toy. Rather, the results of its computations were exceedingly important to all the ancients, including the Romans, where the ill-fated shipment was headed. After all, farmers need to plan plantings and harvests, and priests need to fix religious festivals. Besides that, no ancient worth her salt would want to be caught unprepared by an eclipse. Figure 8.1 shows a beautiful working reconstruction built by Mogi Vicentini. In its original wooden case to hide the inner workings (instead of the transparent plastic), the device would have seemed quite wondrous in the second century BC.[3]

Embedded in the Antikythera mechanism of intermeshing gears is the most remarkable feature of all, and a technological breakthrough: the *differential.* Figure 8.2 shows the most basic form, using a pulley. The pulley is kept in the position $(a+b)/2$, the *average* of a and b, distances to reference points on the two "input" ropes that support it. Examples: If point a moves up and b moves down the same distance, the pulley remains at the same position; the change in b is the negative of the change in a, and the changes cancel out. If points a and b move by the same amount in the same direction, the pulley also moves, by that same amount. If a is held fixed and b moves a certain amount, the pulley and its attached rope move by half that amount.[4]

As a practical matter, the pulley in the differential is usually replaced by a circular gear, and the rope by gears at the left and right, so it is *shaft angles* that are added and subtracted. This is

FIGURE 8.1. One of several working reconstructions of the Antikythera mechanism, this one built by Mogi Vicentini. (Image from Wikimedia Commons.)

exactly how the Antikythera mechanism used the differential, and it is also how your automobile transmits power from the engine to left and right wheels to accommodate their turning at different rates around curves. The differential was independently reinvented many times in the two millennia following its incorporation in the Antikythera astronomical calculator. From our point of view it is an analog adding machine, and we will see it again below in late nineteenth- and early twentieth-century analog computers.

A homesick Richard Feynman wrote back to his family from Athens in 1980 or 1981 after visiting the archaeological museum: "I saw so much stuff my feet began to hurt. I got all mixed up—things are not labeled well. Also, it was slightly boring because we have seen so much of that stuff before. Except for one thing: among all those art objects there was one thing so entirely different and strange that it is nearly impossible."[5] He is referring, of course, to the museum's Item 15087, what has survived of the Antikythera mechanism.

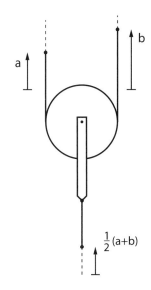

FIGURE 8.2. The simplest form of a differential, a key innovation in the Antikythera mechanism. It is essentially an analog computer that adds or subtracts two quantities, in this case the positions of reference points *a* and *b* on two ropes around a pulley. (After Bromley (1990).)

8.2 More Ingenious Devices

The slide rule

Need I mention the slide rule? Now collectors' items, slide rules were the ubiquitous calculating tools of engineers in the mid-twentieth century, the symbol of the profession, and by far the most widely used analog computer in history.[6] As an undergraduate, your author wore a wooden Keuffel & Esser in a garish orange-leather belt holster, much as today's students lug around laptops with ornamental stickers.

For the many younger readers who don't know what I'm talking about, the slide rule is usually constructed with three strips of wood, plastic, or metal, graduated logarithmically, with one sandwiched between the other two so it can slide between them. Numbers set on one strip can then be multiplied or divided by another by sliding the inner strip with respect to the other two, because adding or subtracting lengths that are proportional to logarithms corresponds to adding or subtracting their logarithms. In fact, the device was invented just a few years after logarithms were invented by Napier in 1614. The slide rule is a much more flexible computing device than the Antikythera mechanism, and the serious engineering slide rule has many scales for calculating not only products and quotients but trigonometric functions and the like.

The Financephalograph

The Financephalograph was invented by Bill Phillips in 1949 and uses water to represent the flow of money in an economy.[7] The machine is also known as the MONIAC, for monetary national income analogue computer. Water is pumped in at the top of the machine, nearly seven feet high, and descends through a central column, where "taxes, savings, and imports are siphoned off into separate loops. Some part of each element rejoins the main flow as government expenditures and the income from private investment and exports. The net flow at the bottom . . . represents the minimum working balances required for a given level of economic activity, and is duly pumped back into the system"[8]

Evidently, about 14 Financephalographs were built and used mainly for teaching. Those were the days before the display screens we now take for granted, and the device was constructed of clear plastic so that the effects of government taxation and spending, consumer spending and savings, and foreign trade could be observed directly. The flow computations performed by the hydraulics were not really prohibitive; it was the visibility of the machine as it operated that made it interesting.

Equation solvers

One problem that comes up again and again in many areas of scientific calculation is that of solving a set of simultaneous linear equations. This is the stuff that algebra homework exercises are made from: "Suppose Judith is 30 years younger than Miriam. How old will they both be when Miriam is twice as old as Judith?" If we let J and M be the ages of Judith and Miriam, we can then set up two conditions: $J = M - 30$ and $M = 2J$. These are two equations, both using only constant multiples of the unknowns J and M, and no squares, cubes, or higher powers—hence the term *linear*. We can substitute M from the second equation into the first, and find that $J = 30$ and $M = 60$. Homework done. By the way, such sets of equations are usually arranged so the constant terms are on the right-hand side: $M - J = 30$ and $M - 2J = 0$, and we assume this below when we refer to the "right-hand sides." This is actually conventional terminology.

Exactly the same sort of problem comes up in designing a truss bridge, say, when we need to calculate the forces on the steel beams to ensure that the bridge supports its load and doesn't collapse. However, the number of unknowns and equations may be not two but easily a hundred. Furthermore, to try different kinds of bridge structures, you need to solve these systems of equations many times with different numbers, and without mechanical calculators (not to mention digital computers) the computational labor becomes enormous.

Sir William Thomson (later Lord Kelvin) is known today for many things, including his contribution to the laying of the first transatlantic cable. This provided the first communication channel between Europe and North America faster than a ship. Being a practical engineer as well as a brilliant physicist, Kelvin asked himself if it might not be worthwhile to design a mechanical device to solve sets of simultaneous linear equations, such as the ones we have just described. He did just that in 1878 and sent off a two-and-a-half-page note to the Royal Society proposing such a machine. Kelvin did not suggest it as a mere curiosity; he notes, "The actual construction of a practically useful machine for calculating as many as eight or ten or more of unknowns from the same number of linear equations does not promise to be either difficult or over-elaborate."[9]

Apparently, nothing more happened along these lines until 60 years later, when John Wilbur, at the time an assistant professor of civil engineering at MIT, took Kelvin at his word. Wilbur was evidently a very serious person; he built a machine out of steel, "with 13,000 parts, weighing a half ton and about the size of a small car" (see figure 8.3).[10] The machine had room for nine equations and, according to Wilbur, could solve for nine unknowns to three significant figures in one to three hours. He compares this with the calculation using a "keyboard calculator," which he estimates "in the neighborhood of eight hours." Today the times sound preposterous, but there was no alternative, and a gain of five or six hours (for each particular case) is highly significant—especially if your computers are human and you are paying them by the hour.[11] Of course the screen on your laptop would barely blink for this task, and the answer would appear almost instantaneously.

A very clear description of Kelvin's machine and how it works is given by Thomas Püttmann, who describes its construction, in great detail, using Fischertechnik, a brand of educational construction kit.[12] Figure 8.4 shows his machine for an example with two equations and two unknowns. Following Kelvin's proposal in principle, there is one loop of string for each equation, and each string passes over one pulley for each unknown. The coefficients in the equation are set by adjusting the position of the pulleys on tilting plates.

This report may strike you as an amusement of historical significance only, but there are important technical points for us here. First, why does it take more than an hour for the machine to find the solution to a nine-variable problem? What is happening all that time? No files to download. No computational loops to cycle through. Kelvin, as we've noted, was an eminently practical man, and he was well aware of the fact that the machine must somehow be brought from an arbitrary initial state to a final equilibrium, from which the values of the unknowns can be read. He states, "The design of a kinematic machine, for success in practice, essentially involves dynamical considerations."[13] In other words, the machine must be given some impetus, and will take some time to come to rest at a useful solution—given that the settings are precise enough and that the friction in the pulleys is small enough. Exactly how this is done depends on the details

FIGURE 8.3. John Wilbur at his machine, ca. 1936. The photo appeared in his paper, Wilbur (1936), and is reproduced here courtesy of the MIT Museum (2011).

of construction, but, one way or another, the machine must be coaxed to its equilibrium state without getting stuck.[14] Wilbur describes a trial-and-error procedure for finding the plate that moves most easily and driving the rest of the machine by rotating that plate.

We must not forget the time that it takes to set the coefficients of the equations into the machine, and the number of these is roughly the square of the number of unknowns. Solving a

FIGURE 8.4. Püttmann's construction of Kelvin's machine for solving simultaneous linear equations, using the construction kit Fischertechnik. The coefficients of the equations are set by sliding the four pulleys on the seesaws at the bottom. (Those seesaws can be seen as the tilting plates in figure 8.3.) The two wide scales at the top show the components of the solution, and the two circular dials at the top left and right set the right-hand sides of the equations. (Courtesy Thomas Püttmann.)

system with nine unknowns involves setting about one hundred coefficients, each setting being made with micrometer screws. It is hard to know from Wilbur's description just how the time breaks down between dialing in the coefficients and actually operating the machine.

How about the three significant figures? What if we need more accuracy? Lord Kelvin also considered this question in 1878, and, typically, his thought processes leaped a century. He pointed out that once one has a rough solution from the machine, say, to one to three figures, one can perform the relatively fast and simple calculation of substituting the rough solution into the original equations to find the differences between the computed and the stipulated right-hand sides. This leads to another problem of exactly the same form, with exactly the same coefficients (but with new right-hand sides), that will tell us how to adjust the old solution to get a new one that is more accurate. This procedure can be repeated to get a solution with any required accuracy. What is important here is that computation of the errors in the right-hand sides (called the *residuals*) does not require the *solution* of the system but only the *substitution* of the current solution into the original equations. Kelvin's insight anticipates the use of *hybrid* computers, which combine digital and analog techniques with the aim of exploiting the advantages of both. The idea gained a certain currency before Moore's law took full effect, but died, perhaps prematurely, by the end of the twentieth century, when digital computers became the only game in town. We revisit the hybrid machine later when we discuss the brain, which, in fact, uses both digital and analog computation.

Insofar as it illustrates some technical points about the speed and accuracy of analog computers, that is all we need to know about Wilbur's machine. However, the *MIT Civil and Environmental Engineering Newsletter* (2001) supplies some tidbits that are suggestive of a mysterious afterlife. First, the newsletter tells us that, "the Wilbur Simultaneous Calculator disappeared without a trace ... after having obstructed the hallway outside of Rm. 1-390 for many years. Nobody at the MIT Museum, Boston Museum of Science, or the Boston Computer Museum has any clue what happened to it." (I remind the reader that the machine was the "size of a small car" and had 30,000 parts.)

Next, the editor of the newsletter (Debbie Levey) reports that she "was astonished to see photos of an almost exact replica of the machine in a magazine article about the Tokyo Museum of Science." Correspondence with science writer Seishi Koizumi "revealed that before World War II the Japanese had duplicated the machine" and put it to use in aeronautical research. The Japanese copy was planned for display at the National Science Museum in Tokyo in 2002. There the trail grows cold, and I end the historical diversion. It is time to return to our examination of analog computation.[15]

8.3 Deeper Questions

I have glossed over two natural questions about the operation of a device of the Kelvin type—questions that come up whenever we try to solve *any* problem, by machine or otherwise: First, what if there is no solution at all? Second, what if there is more than one solution? As we shall see, these two questions are not incidental details; they play a central role in our attempts to understand the power and limitations of computing in general.

In the case of simultaneous linear equations, we can easily see how these situations can come about. For example, two equations can be contradictory. In our Judith and Miriam problem, there is nothing to prevent us from specifying that $M - J = 30$ and $M - J = 31$. Clearly, there can then be no solution, no matter what the rest of the problem stipulates.

Another possibility is that two equations can be redundant. For example, we might stipulate that $M - J = 30$ and $2M - 2J = 60$. The second equation is just the first with every term multiplied by 2. It tells us nothing new. Mathematical theory tells us that in this case there will be many solutions, in fact, an infinite number of them.

As we might expect, attempts to solve such contradictory or redundant sets of equations will exhibit symptoms that something is wrong. The symptoms will depend on which particular machine we are using. Easiest to discuss is the construction-kit machine in Püttmann (2014), shown in figure 8.4. The way problems are set up on this machine, with one string for each equation, the strings are initially loose and are sequentially tightened by pressing on the panels that determine the different unknowns. The initial loose condition means that the equations we want to solve are not strict but have slack between their left- and right-hand sides. When there is a solution to the original set of equations, we are driving the state of the machine to a point where all the equations are satisfied simultaneously.

In the case when the original equations are incompatible, we will reach a point where an equation (technically, a *constraint*) is loose and cannot be tightened. The solution process will simply get stuck.

In the case when there are many solutions, the behavior is even simpler. We will be able to find a solution by the usual procedure, but the point we reach will in general depend on our starting point, and, furthermore, when we do reach a solution, it will be "loose" in the sense that we can still slide the variables through a continuum of valid solutions.

We can take away an important lesson from our consideration of Kelvin's equation-solving machines: mathematical difficulties that are inherent in a problem will manifest themselves, one way or another, when we try to solve it. If we build an analog machine for the problem, we can expect the manifestations to be physical. The machine might get stuck or reach a slippery solution. If we use a digital computer, we can expect corresponding numerical symptoms, like attempted division by zero, which is exactly what happens in the simultaneous equations problem. Mother Nature does not allow us to sidestep essential difficulties; physics and logic tap the same reality.

8.4 Computing with Soap Films

We now step across an important threshold, from problems we know very well how to solve, to problems that have defied the best attempts of even the cleverest of scientists.

As mentioned at the beginning of this chapter, the earliest analog and digital computers were viewed as problem-solving tools. Moore's law and the resulting personal computer, with the help of enormous market forces, shifted attention to the digital machine as, essentially, a data processing or, perhaps more kindly, a digital signal processing machine. Today computer scientists work to make computers faster, more reliable, more secure, smaller, and cheaper. We now know more or less how to make excellent smartphones, laptops, cameras, and all the other gadgets that make modern life, well, what it is.

Scientists, who built the first large and clumsy machines to solve their computational problems, have, with some irony perhaps, been free riders on this technological train, happily burning cycles on their desktops for their problems and, furthermore, sharing their research with colleagues across the globe in seconds. From our perspective, however, where we ask why the world became digital, we have come full circle. We have returned to viewing the computer as a problem-solving tool. The reason is simple and compelling: we want to know what is possible and what is not, whether we might be overlooking unique resources in sticking with digital machines. In short, we must plan for a future when problem solving, and not signal processing, is the dominant intellectual challenge.

Here is a very pretty little problem, called *Steiner's problem.* Suppose we are given N towns on a flat Earth, and we wish to connect them with a network of roads. How do we do so with the least possible total length of roadway? The problem, or at least the simple three-city version, can be traced back to Fermat, in the seventeenth century. However, the name "Steiner's problem," after the nineteenth-century mathematician Jakob Steiner, has stuck.[16]

As a simple example, if $N = 3$ and the three towns are at the corners of a regular (equilateral) triangle with sides of length 10 miles, say, the solution is to choose the center of the triangle as a junction point and connect the junction to each of the three corners with straight roads. The most obvious alternative is to connect one city to the other two with straight roads (without a junction point), but that solution has a total road length of 20 miles, whereas the "star" solution has a total length of 17.32 miles. This represents a savings in concrete of about 13%, assuming that, in our ideal world, we aren't considering the added expense of a traffic light at the three-way intersection of the star solution.

The real difficulty in this problem, which becomes apparent when the number of towns grows much beyond three or four, is the embarrassingly large number of possibilities that arise for the choice of junction points—which, by the way, are called *Steiner points.* We do have some help from the mathematicians, who have proved that we never need more than $N - 2$ Steiner points, and that at any Steiner point exactly three roads meet, always at angles of 120°. Beyond that, however, we are faced with an exasperatingly large number of choices: How many Steiner points? Where do they go? And which towns get connected to which? Figure 8.5 shows some candidate solutions in an example where the towns are situated at the corners of a regular hexagon. The variety of solutions for even this tiny example begins to show how complicated the choices can become. Try to imagine the profusion of choices when we have, say, a hundred towns that are not regularly placed.

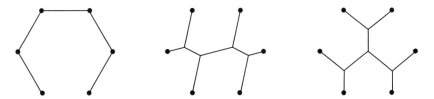

FIGURE 8.5. Three candidates for a solution to the six-city Steiner's problem where the towns are at the corners of a regular hexagon. (After Isenberg (1976).)

The classic book by Courant and Robbins popularized the idea that Steiner's problem can be solved by dipping a wire frame in a soap solution and withdrawing it, leaving a soap film that reveals the solution.[17] Belgian physicist Joseph Plateau had done extensive experiments with soap films in the 1870s, and the mathematical study of minimum-surface-area problems goes back a hundred years before that, to Euler and Lagrange. The physical intuition behind the method is that a soap film tends to form a surface of smallest possible area, because that minimizes the potential energy due to surface tension. For example, dipping a circular wire frame will, on withdrawal, yield what we expect, a film that is simply a flat disk bounded by the wire frame. If it weren't flat it would have larger area.

Figure 8.6 shows how a Steiner problem can be set up with wires between two parallel plates. The wires are perpendicular to the two plates and run between positions on the plates corresponding to the locations of the cities. A surface of minimum area will then consist of flat sheets that form between the plates, and their trace on one plate or the other will show us where the roads must go in a solution of minimum total length. It appears that the physics, via our saponaceous little analog computer, has circumvented the problem of having to consider an unmanageable number of possible choices for the number and locations of Steiner points. Or has it?

8.5 Local and Global

If we think that the soap-film computer has triumphed, repeating the dipping experiment many times on any problem that is

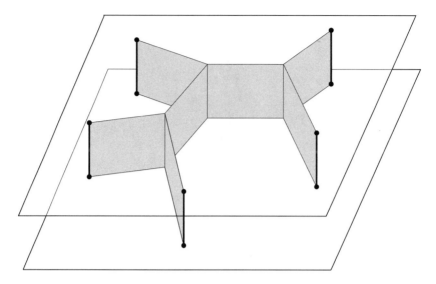

FIGURE 8.6. Solving a five-city example of Steiner's problem by dipping a construction in a soap solution, withdrawing it, and observing the resulting soap film. (After Courant and Robbins (1996), p. 392.)

complicated enough will quickly disillusion us. We will find that different solutions will appear, with different arrangements of Steiner points and even different numbers of them. We will still be left with the problem of deciding which of the myriad candidate solutions offered to us by the soap films is the best. Where has the reasoning gone wrong?

The resolution of the apparent paradox reveals something very important about the underlying structure of this and many other problems. The solutions offered by the soap films do not actually have *minimum* surface area but only *locally* minimum surface area. That is, each solution cannot be improved by perturbing it a little bit; each solution is best in a *neighborhood* of candidates that are close by. But the film cannot make a big jump to a solution of an entirely different character—it doesn't see the big, global picture.

Searching for a soap film of least surface area is a rather abstract problem and is difficult to visualize. A concrete way of looking at it is to imagine instead that you are exploring a craggy landscape of peaks and valleys, looking for the lowest point. In this picture, your altitude represents the surface area, which we want to be minimum, and your geographic location represents the choice of a configuration of the soap film. At some point in your exploration, at the bottom of a valley, you might find yourself quite convinced that you have reached the ultimate lowest point. Everywhere you look, the ground is higher. But of course, there might be another valley, beyond your range of vision, that is lower. You have no way of discovering that without leaving your valley, which might be an excursion that is too risky or time-consuming.

This situation comes up all the time when computer scientists study problem-solving techniques; in general, more information about the problem itself is needed to guarantee that a particular valley is actually the lowest possible. The terminology for this phenomenon is that the short-sighted best is a *local* minimum as opposed to a *global* minimum.

In the same way, the physical principle of least surface tension can only find improvements of a solution that are in its neighborhood, reachable by incremental variations in the current solution. The soap-film analog computer finds only local minima, a fact that is confirmed by damp experimentation.[18]

As in the matter of solving simultaneous equations, mathematical difficulties have manifested themselves physically when we've tried to use analog computers—but in a much more dramatic, perhaps disturbing, way. The intuitive notion that there is some sort of essential difficulty connected with certain problems is a central theme in what is coming soon, the study of the difficulty of problems in general.

Before we go on to present-day and digital manifestations of the "essential difficulty" that we have hinted at, we describe, briefly, the progression of analog computers from special-purpose, mechanical contrivances to their present state, which many might regard, perhaps mistakenly, as a dead end.

8.6 Differential Equations

The analog computers we have mentioned up to now solved specialized problems and hence only very particular equations. As another example, wind tunnels have been used for more than a hundred years to study the flow of air around objects like airplanes, automobiles, and buildings. In effect, they solve the equations of fluid dynamics, but only those. A scale model of the airplane, say, is built, and air is blown past it. Measurements then reveal the aerodynamic forces on the airframe. Wind tunnels as analog computers have been very useful, especially before digital computers were available, and they are still used to confirm the results of numerical digital computation. However, they only solve problems of fluid dynamics. If you have invested in a very expensive wind tunnel and one day you want to study the formation of galaxies in the early universe, you are out of luck. You need to build an entirely different computer.

The key advance was building an analog machine that can solve any problem in the very broad class described by *differential equations.* Differential equations are the bread and butter of the engineer and physical scientist, and are used to describe almost any kind of problem you can think of: the motion of the planets and stars, the competition between species in a competitive environment, the flow of current in electrical circuits, of heat in materials, of electrons in semiconductors—the list goes on forever. A standard way of doing science is to formulate a differential equation that describes your problem and then solve it—with luck using pencil and paper in terms of well-known functions, but probably, if the system is at all new and interesting, using a computer. Differential equations involve the *rates of change* of physical quantities.[19]

One of the simplest examples arises when we model the growth of a particular organism, say, bacteria, in a nutrient medium. The most naïve view is that the more bacteria there are, the faster the number of bacteria, say, N, increases. There are just that many more bacteria to reproduce. The first differential equation we might write would state that the rate of change of N is proportional to N. This leads to a solution where N increases exponentially, as we expect, since there is no limit to growth. Thomas Malthus wrote about this at the end of the eighteenth century, and today we refer to the resulting, terrifying exponential curve of population growth as the *Malthusian law.* Pierre Verhulst took the next step about 40 years later, when he modified the differential equation to take into account the fact that the growth rate of bacteria is limited by competition for nutrients. He did this by adding a factor that decreases as N increases. That leads to a more realistic and very successful solution, where the number of bacteria starts out following the Malthusian law but then levels off to what is called the *carrying capacity* of the medium. This is just one example of how biologists, and most other scientists, study systems that interest them, proceeding through successively refined differential equations.

8.7 Integration

It may now seem that solving a differential equation is simply a matter of interconnecting devices that find the derivatives of signals. Not so, and for a reason that takes us back to the inevitable limiting factor that runs through our narrative like a red thread: noise. Measuring the rate of change of a physical variable is an inherently noisy process. For example, when you ride on an uneven road, you feel the bumps where the road changes elevation suddenly, not the general level of the road. For this reason, general-purpose analog computers always work with the opposite of the differentiator, the *integrator*. The rate of change of the road height may be very large at a bump, while the actual change in road height may be very small.

The process of integration reverses the process: given the bumps in the road (its derivative), we find the actual elevation by *smoothing* (*integrating*) the derivative. We can also think of the integrator as *accumulating* a rolling sum of the bumps. Integration is such an important tool of physicists and engineers, and is such a fundamental idea in its own right, that it has attracted the attention of the best mathematicians, including Archimedes (implicitly) in the third century BC and Leibniz and Newton in the seventeenth century. We won't find it necessary to embark on a short course in calculus here. It is enough to have the intuition that integration is a smoothing operation, and the opposite of differentiation, which is noisy.

If I may interpose a bit of nostalgia: I recall sitting in the local library as a young science buff, searching through books for the secrets of the universe. A strange approach today, perhaps, but there were no search engines then, or computers for that matter. It seemed that every time I tried to go beyond Gamow (1947) and similar popularizations, I would encounter the mysterious mathematical sign for integration: a tall, slim, and somewhat archaic "S," which began to take on an aura of forbidden knowledge. I dare not print it here. It is best to go on.

To see how a differential equation can be solved on an analog computer using integration, consider the simplest case, the Malthusian equation above: The rate of change of the number of bacteria (the derivative of N) is proportional to the number of bacteria (N). The more there are, the faster their numbers increase. The important observation is that if the derivative of N is proportional to N, then the *integrals* of the derivative of N and of N must also be proportional. The "integral of the derivative of N" is, by definition, just N (integration is the opposite of differentiation, and the two operations cancel out). So we have a new equation that is equivalent to the original: the integral of N is proportional to N. It is now easy to set this up if we have, say, a mechanical integrator (coming soon). We just take its output, multiply by a constant (the proportionality constant), and couple it, mechanically or electrically, to its input. Solving any differential equation on any kind of general-purpose analog computer proceeds essentially along these lines.

8.8 Lord Kelvin's Research Program

Building a machine that can solve any differential equation quickly and accurately is a revolutionary idea with far-reaching consequences. Once again, it was Lord Kelvin who saw the potential of a machine solution and took the first and important steps in this direction. And again, the work was picked up 50 years later at MIT, this time by Vannevar Bush.

Looking back at Appendix B' of Thomson and Tait (1890), which is a carefully selected sequence of certain papers in the *Proceedings of the Royal Society* in the 1870s, it is clear that Kelvin had laid out a sharply directed research program for himself with the goal of mechanizing computation. I outline these here, preserving their Roman numeration:

I. The first paper describes a tide-predicting machine, which performs Fourier *synthesis*, adding together the different frequency components of the tide produced by the Moon and Sun. The operation of Fourier synthesis is the opposite of Fourier *analysis,* which we've already used in chapters 2 and 6, and the two operations are a closely related pair, each undoing the other. Their applications in the sciences and technology are ubiquitous. For example, you probably use Fourier analysis many times a day; it is at the heart of the JPEG image-coding format. As another example, the earliest experiments in computer music performed sound synthesis from Fourier components, exactly the same computation as Kelvin's tide-predicting machine.

II. Then comes a description of Kelvin's machine for solving simultaneous linear equations, which we have seen was later implemented in earnest by Wilbur.

III. The next paper was actually written by James Thomson, Kelvin's older brother and frequent collaborator. It describes an improved mechanical integrator, the heart of the machine designed to solve differential equations, which came to be known later as the *differential analyzer*. The basic idea is inherited from the *wheel-and-disk integrator*, which was developed from the *planimeter*, a mechanical device for finding the area of a figure traced on paper.[20] A very simplified sketch of the wheel-and-disk integrator is shown in figure 8.7. A wheel acts as a platform, and a disk rests on it and rolls at a varying distance $f(x)$ from the wheel's center, where x is the angular position of the wheel.[21] The rotation of the little disk, and hence its angular position, integrates the varying distance $f(x)$.

The problem with the wheel-and-disk integrator is that the disk needs to slide as well as roll, which bothered Kelvin a great deal. His brother James Thomson invented an improved version, which uses a ball resting between the wheel and a recording cylinder, and Kelvin adopted the wheel-ball-cylinder integrator for his machines.

IV. Kelvin then describes how to build, using the integrator in Part III, a *Fourier analyzer.* As mentioned, this machine

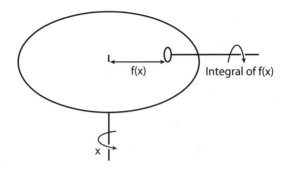

FIGURE 8.7. The principle of the wheel-and-disk integrator. As the wheel is turned through an angle x, the disk rolls along it, at a distance $f(x)$ from its center, and thus accumulates the integral of $f(x)$. (After Irwin (2013/2014).)

        performs the operation opposite to that performed by the Fourier synthesizer, the tide predictor in Part I.

V. He proposes machines for solving differential equations with second derivatives (second-order).

VI. Then the same for differential equations of any order.

VII. Kelvin then describes the actual construction of the machine proposed in Part IV, which he calls a *harmonic analyzer*, for analyzing tides. The machine is now on display at the Science Museum, London.

It looks as if there was no stopping Kelvin, but it was here that he ran into a roadblock that took 50 years of technological development to overcome. Kelvin's problem was that he had no way of transmitting the results of one intermediate calculation to several others. The information in mechanical form was weak and became ever weaker as it passed from one stage of analog computation to the next. The same problem arises in the electronic logic circuits of digital computers, where the output of a gate needs to be passed to several other gates. The general process is called *fan-out*, and in electronic computers the fan-out problem is solved with electronic amplification. The output of each stage must be strong enough to drive all the other stages it is connected to. It was not until 1925 that Henry W. Nieman invented the mechanical counterpart of an electronic amplifier, his *torque amplifier*, and this is the missing piece that Bush put to work in realizing Kelvin's quite general differential equation solver. From there the path to a general, mechanical analog machine was clear. One implementation of it is described in Bush (1931).

Vannevar Bush was one of the most influential scientists of the twentieth century. He worked with Wilbur on his linear equation solver, built the differential analyzer just described, and went on to oversee many important projects during World War II, including the establishment of the Manhattan Project. He also mentored Claude Shannon as a graduate student at MIT, the Claude Shannon whom we met as the founder of information theory. In 1941 Shannon wrote a fundamental paper about Bush's analog computer, titled "Mathematical Theory of the Differential Analyzer," in which he proves that a very wide class of differential equations can be solved using a differential analyzer of the type we have been describing.[22]

8.9 The Electronic Analog Computer

It was natural that mechanical analog computers, which had to be built in the machine shop, were replaced by electronic versions of the same idea, with electronic counterparts of integrators, adders, and scale constants. And as we know, Moore's law and digital computation soon wiped out the resulting general-purpose electronic analog computers. However, there was a time window in the 1950s and 1960s when commercial analog and digital computers competed neck and neck, and part of my own undergraduate education entailed programming an analog computer not unlike the one shown in figure 8.8. "Programming" in this case meant wiring the *patch panel*, the switchboard where the different components could be interconnected with wires that plugged into the panel. In figure 8.8 the patch panel is featured in the foreground, and you can see why I referred to it as a "rat's nest" in chapter 1.

These days, if a particular differential equation is not one of a very few that have known solutions, a digital computer is used to solve it numerically, and the required numerical arts for doing so are highly developed and readily available in scientific computing packages. Digital computers are so fast today that there is rarely, if ever, any consideration given to using an analog machine. But

FIGURE 8.8. "Computers Speed Aircraft Design"—an electronic analog computer (foreground) during the competitive years: a 3 × 15 array of vacuum tubes sits atop the computer console. (The cover of *Radio-Electronics Magazine*, January 1959; available at http://www.americanradiohistory.com/. Accessed September 15, 2017.)

when digital computers were in their infancy, it was sometimes much easier to throw together a circuit on the patch panel than write code—for a slow and expensive digital machine, without a screen editor and, for that matter, without even a compiler!

The demise of the electronic analog computer brings us to what is evidently the end of analog computation and the beginning of the digital computer era. We can now return to a broader view of the computer as a problem-solving machine and the idea, introduced at the beginning of this chapter, of the "essential" difficulty of a problem. We describe how present-day computer scientists think of computation, and that thinking is almost always in terms of discrete machines. I've hinted that funeral services for analog machines might, conceivably, be premature, but we must postpone that question as we change the scenery, the cast, and begin a new act.

9 Turing's Machine

9.1 The Ingredients of a Turing Machine

Today the bare term *computer* implies digital, not analog, and *computer scientists* are almost always studying completely digital machines. Computer theoreticians, in fact, work with the perfectly discrete picture of a computer invented by Alan Turing about 80 years ago, and we discuss their main conclusions (and conjectures) about the power of such a machine—the *Turing machine*—in the next chapter. My aim in this chapter is to construct the Turing machine from scratch using two very basic principles that were discovered, astonishingly, in the early nineteenth century. The first principle is, to use the modern term, the *stored program*, perfected and brought into practical use in the weaving industry by the Frenchman Joseph-Marie Jacquard, and the second is *branching*, or *conditional execution*, conceived by the Englishman Charles Babbage.

The recipe for a Turing machine is (1) use information only in discrete form, (2) isolate the control as a stored program, and (3) make provision for the execution of the stored program to depend on the results of previous calculations. It took two centuries for technology and mathematics to catch up with Jacquard and Babbage.

We now take the stored program for granted; what present-day all-digital computers do, step-by-step, is determined by a *program*, or sequence of *instructions*, written in some language that is convenient for the programmer, but translated into a more basic language that is directly interpretable by the machine's hardware—the interconnected gates described earlier. In all-digital machines it is thus easy to distinguish between the control, which appears as code of one kind or another, and the computation itself, which is directed by that code and takes place in the special little chip called the central processing unit (CPU). But how does this distinction between control and computation work out for machines that are all or partly analog? We illustrate some answers with examples in the following section, starting with the opposite of the digital computer, the all-analog machine.

By the way, in a modern digital computer the data on which the program operates is usually—and carefully—kept well apart from the program. However, the instructions themselves can, theoretically, be treated as data, and programs that operate on their own code are called *self-modifying*. Such programs are usually considered dangerous, both because they are tricky to get right and because they invite attacks from clever programmers with evil intent.

9.2 The All-Analog Machine

For a particularly simple and clear-cut example of a purely analog machine, return to the Antikythera mechanism discussed in the previous chapter. It is entirely and unmistakably analog. The machine is a collection of meshed gears, the results are presented with indicating dials, and the rotations of the gears and shafts— in other words, all the moving parts—are free to vary continuously. The operation of the mechanism is innocent of any discrete activities.

Consider now how we might separate the control and computation in the Antikythera mechanism. The "program" is not written out in what we ordinarily think of as a language. Rather, it is embodied in the choices of which gear meshes with which, and the ratios of teeth on those gears. The control of the computation is "structural," embodied in the way the mechanism is put together. But the computation performed by these gears, predicting the position of the planets, Moon, and the occurrence of eclipses, is carried out by these very same gears. The control and computation as well as the data—the rotational position of every gear—are all tightly intertwined and not at all separable.

9.3 The Partly Digital Computer

The wristwatch

The Antikythera mechanism is a logical precursor to the clock, and, in fact, the work of the Antikythera craftsmen was picked up, at least in spirit, by early clockmakers, who started producing clocklike heliocentric models of the solar system called *orreries*. These are often very beautiful and ingenious mechanisms in their own right, but I mention them because it is the mechanical clock that I want to discuss next.

These days, a high-quality, totally mechanical spring-powered wristwatch is a prestigious possession, and can easily cost a thousand times as much as a digital watch, which counts the oscillations of a quartz crystal and reduces the count to a digital display. We usually refer to the former as an "analog watch" and the latter as digital—just another electronic gadget. But not so fast! If we examine the operation of a so-called analog watch, we find an irreducibly discrete element: the pallet that rocks back and forth, engaging first one tooth of the escape wheel and then another. Together, the pallet and escape wheel are controlled by the oscillation of the balance wheel, and they also keep the balance wheel going by transmitting impulses from the mainspring.

The hands of a mechanical wristwatch move in discrete steps, apparent if you examine the second hand with a magnifying glass. The mechanical wristwatch therefore has both analog and digital aspects, not so easily separated. And, as in other gear-based mechanisms, the control and computation are both coded in the structure and are hardly separable.

The gears in the Antikythera mechanism move *continuously*, but the gears in a watch are *latched*—they jump when the pallet rocks, and they can be considered discrete components. The time it takes for the balance wheel to complete a basic oscillation is determined by the free, analog motion of the balance spring; the mechanical watch can therefore be thought of as an *analog-controlled digital computer*. We don't have far to look for the exact opposite situation.

The electronic analog computer

Consider the patch panel, or rat's nest, used to interconnect different parts of the electronic analog computer. As mentioned before, the patch panel works the same way as a telephone switchboard: There is an array of sockets, and the operation of the analog computer is controlled by plugging cables into pairs of sockets, establishing electrical connections between certain components. Programming consists of deciding which analog components get connected to which, and then actually connecting them electrically. These connections are *digital* in the sense we use the term: each possible connection is either ON or OFF. But the resulting flows of current are continuously variable and subject to noise, and are *analog* in the sense we have discussed earlier at some length. Thus, the general-purpose electronic analog computer, described in the previous chapter, is really a *digitally controlled analog machine*—the opposite of the watch. The control and computation are separated but not perfectly, because it is still the case that the cables that interconnect the components play two roles. We are getting there, however.

9.4 A Reminiscence: The Stored-Program Loom in New Jersey

In the 1940s the hot, humid summer nights of my childhood in northern New Jersey were filled with the soporific *chug-a-chug-a-chug-a* of the massive embroidery machines in the neighborhood factories. During the war the machines ran 24 hours a day, weaving insignias for soldiers half a world away. On blazing, glorious school-less afternoons, I would climb in an open window from a garage roof and watch the "watchers," who scanned the marching rows of brightly colored patches for signs of broken threads, which they would deftly set right as the machines continued to run.[1]

The machines are properly called *schiffli embroidery machines*, because the shiny steel shuttle that holds the thread for the lock stitch is shaped like a miniature ship's hull; *schiffli* is German-Swiss for "little ship." I certainly didn't know it at the time, but the schiffli embroidery machine is a direct descendant of the Jacquard loom, invented soon after the French Revolution and a close cousin of the rudimentary digital computer that was being developed by Turing and his coworkers in England to decipher Nazi codes.

The row of needles that execute the embroidery pattern is controlled by holes on a paper tape, and there, precisely, is the crucial, complete isolation of the stored program that we are hunting down. The holes on the punched paper tape are the program—and the dance of needles the execution.

9.5 Monsieur Jacquard's Loom

Figure 9.1 shows the French weaver and inventor Joseph-Marie Jacquard seated in his workshop. We appear to have interrupted him at his work, and he seems to disapprove. He holds a pair of dividers in his right hand, and the key to the image is the pile of holed cards beneath. These are strung together to control his automatic loom, as shown in the miniature model behind.

In operation, an array of needles is pressed against each card in turn, and those needles that encounter holes move hooks that lift the warp threads that correspond to the holes in the card. In this way, the holes in the card control, for each warp thread, whether the perpendicular weft thread passes under or over that warp thread. The embroidered design is sewn one row at a time, each row controlled by enough cards to cover the number of warp threads in the loom.

It is easy to mistake the image of figure 9.1 for an engraving, but it is actually a tapestry woven from black and white silk on a Jacquard loom. The "program" used 24,000 punched cards, many more than usual for the production of the usual fashion fabrics. The old Jacquard cards appear to have had 6 rows and 8 columns of possible holes, 48 bits per card, and the cards were strung together to be fed to the loom. If I interpret the figure of 24,000 cards correctly as meaning 24,000 of these small cards, this means that the portrait of Jacquard represents what we think of today as 144 KB. Jacquard was a man of great vision, but I doubt that he anticipated the invention of image compression.

FIGURE 9.1. A woven silk portrait of Joseph-Marie Jacquard, produced on the loom of his own invention. It is based on a painting by Claude Bonnefond that was commissioned by the city of Lyon in 1831. What appears to be a bullet hole in the window may be a sly reference to the fierce and sometimes violent resistance from silk weavers to the introduction of automation in the weaving industry. (Courtesy the Metropolitan Museum of Art; available at http://www .metmuseum.org/art/collection/search/222531. Accessed January 22, 2018.)

With the typical high-quality JPEG compression ratio of ten-to-one, however, this image corresponds to something like a 14 KB black-and-white image, and the general appearance of the portrait seems about right. It's hard to imagine today, but the analog-to-digital conversion from the original oil painting was done by hand, pixel by pixel.

Jacquard's loom met swift and wide approval. It was patented in 1804; by 1812 there were about 11,000 in operation in France, and by 1832 about 800 in England.[2] The invention, however, did not receive the same enthusiastic reception from the silk weavers of France and England. It was, after all, about 24 times faster than a manually operated drawloom, and a weaver could operate it alone, without the assistance of a draw boy. Workers in the weaving industry quite understandably perceived Jacquard's loom as a threat to their employment, and it played a role in the anti-technology Luddite movement in England.

Jacquard was not the first to try to use punched cards or paper tape to control a loom in one way or another. He borrowed ideas from several predecessors whose machines were only partly successful. But his apparatus for controlling a powered loom was the first that was automatic, reliable, and fast.[3] It was a breakthrough. Perforations in paper or cardboard for recording information had a history stretching back at least to the eighteenth century. The idea has since been used in player pianos, and then very widely by small computers from the 1960s through the 1980s. I've already mentioned, in chapter 2, Godfrey Winham's digital-to-analog conversion program, written in the early 1970s, which was always on hand in Princeton's computer-music lab for loading from paper (or mylar) tape. Given the advances in modern magnetic and electronic storage media, paper tape is long obsolete, with the possible exception of situations where its advantages are still decisive: it is immune to electromagnetic fields; readable, in a pinch, by eye; and quickly and easily destroyed—ideal for military and spy work.

9.6 Charles Babbage

We are not finished setting the stage for Alan Turing. The idea of a stored program is very important, but it is not enough in itself to build a useful computer. A critical piece is missing: the program of a Jacquard loom will keep producing the same pattern, complex though it may be. You will never be able to program it to serve as a web browser, read email, or even just sort a list of names. The missing piece is the ability of the machine to decide what it is going to do next *by examining the results of previous operations*. The idea is familiar to anyone who has written a program that uses what is called a *conditional statement*, such as an "if" or "while." This new, key element was supplied by Charles Babbage, who combined it with the idea of a stored program; with that he can properly be said to have invented, in his way and for his time, the digital computer—plain and simple.

Charles Babbage was born in 1791 and began his professional life as a mathematician, publishing more than a dozen papers between 1813 and 1820, and establishing himself as a respected researcher. Babbage's own list of his publications is reproduced in Campbell-Kelly's edition of Babbage's entertaining memoirs.[4] Item 18 of that list reveals a sudden departure from his traditional mathematical interests to the subject that would absorb most of Babbage's attention and energy for the rest of his life: it is titled, "Note respecting the application of machinery to the calculation of mathematical tables," and was published in the *Memoirs of the Astronomical Society* in 1822. Campbell-Kelly, editor of his collected works, attributes this idea of mechanized calculation to his collaboration on a set of astronomical tables with his friend, the famous astronomer John Herschel, who had been his classmate at Cambridge:[5]

> In the course of our conversations on the subject it was suggested by one of us, in a manner which certainly at the time was not altogether serious, that it would be extremely convenient if a steam-engine could be contrived to execute calculations for us; to which it was replied that such a thing was quite possible, a sentiment in which we both entirely concurred and here the conversation terminated.

At this point Babbage embarked on the planning of what he called *Difference Engine No. 1*, so called because it relied only on the operations of addition and subtraction.[6] This, and his future plans for No. 2, carry out fixed programs and are of interest to us here only because they served as a springboard and inspiration for his *analytical engine*. In all his work on calculating machines, Babbage played the roles of theoretician, designing the machines; engineer and technician, building the machines; and grantsman, seeking government funding for his projects.

Babbage was a man of some complexity. On the one hand, he was the brilliant and often charming host of regular Saturday soirées at his home, attended by the cream of the intellectual life of Victorian London, including, of special interest to us, his friend Sir John Herschel and Lord Byron's daughter Ada Lovelace.[7] On the other hand, he could be both irascible and naïve, and was generally a poor manager. His dealings with his government sources of funding were frustrating, especially those with Prime Minister Sir Robert Peel, and his support eventually dried up.

Popular accounts of Babbage frequently remark that he was never able to complete any of the machines he envisioned—although he did finish "one-seventh of Difference Engine No. 1, a demonstration piece consisting of about 2,000 parts assembled in 1832...[which works] impeccably to this day."[8] Indeed, he carried on a continual struggle to realize working models of his machines, designing special machine tools, pushing the limits of contemporary fabrication techniques, working up a profusion of detailed engineering drawings, and struggling with the considerable funding demands. In those days it was mechanical computation powered by a steam engine, or nothing: electricity was poorly understood at the time, and he did not have the option of trying out ideas with cheap electronics. Perhaps Babbage's greatest problem was his fixation on the goal of translating his abstract ideas to steel and steam: his imagination would often outrun his plans for practical construction. If Charles Babbage led a frustrating life, it was in large part because he was trapped in the wrong century.

At least for our purposes, I suggest we think of Babbage as the first theoretical computer scientist and leave aside his fascinating, if unfinished, multiton computing engines. In a way, they were his pencil and paper. He thought about computation in the most concrete way, quite reasonably for someone born when

George Washington was president of the United States. From the perspective of computer science, then, his marvelous contribution was, as mentioned above, to combine conditional execution with a stored program, and so to describe the first truly *general* digital computer. We have more to say about exactly what this important word "general" means after we discuss Turing. But we turn now to the heart of his creative thinking, his analytical engine.

9.7 Babbage's Analytical Engine

The many, often complex ideas required for the analytical engine were in a constant state of flux, and Babbage was continually simplifying and refining them. He was also a man who evidently lacked the discipline to leave full descriptions of the state of the conceptual machinery at any given time, and the result is that he himself left only general, often frustratingly vague descriptions of the many versions of his analytical engine. We next summarize, briefly, the many new ideas incorporated in those machines, all of which have counterparts in today's modern computer.[9]

The stored program: Babbage greatly admired Jacquard's portrait-in-silk and managed to obtain his own copy, despite the fact that they were made only to special order and not at all widely available. Each copy took many hours to produce, even on Jacquard's automatic loom. His notes show that he quite consciously borrowed Jacquard's idea of using punched cards to control his machine, in June of 1836. These replaced an awkward system using drums with studs, and had the important advantages of eliminating errors in setting the studs in the machine and allowing programs of unlimited length.

Sequential programming: The sequence of operations on the "repeating apparatus" of the analytical engine was to be controlled by what Babbage called "combinatorial cards," or "operational cards," eerily reminiscent of the decks of 80-column cards used to store programs from the 1960s into the 1980s—then ubiquitous, now obsolete. The instructions on these cards could control *branching*—the conditional execution, depending on previous results, that, as we have said, is so important for putting together a machine capable of general computation. Also remarkable is the provision for feeding the output of the computation back to its input, which Babbage described as "the engine eating its own tail." This allows iterating sequences of operations, or what we call *looping*.

Multiplication and division: The difference engines required only addition and subtraction (which are practically the same thing in terms of hardware), but Babbage wanted more operations for his envisioned general computation. He subsequently arrived at a suite of the four basic operations of addition, subtraction, multiplication, and division, just what we expect from a modern CPU.

Separation of processing and memory: The data in the analytical engine is in what we would today call memory, which Babbage called the "Store," and was separate, at least conceptually, from the central processing unit, which he called the "Mill."

Printing: Provision had already been made for printing the results of the difference engines, even envisioning the creation of plates for mass printing. Errors were inevitably creeping into hand-calculated tables, and error-free printing was very important to Babbage. He even planned a "curve-drawing apparatus." These were what we might today call peripherals, and could even operate offline from the punched-card output of previous calculations.

Efficiency and computation speed: Babbage paid a lot of attention to the time that his arithmetical operations would take, very much in the spirit of the modern complexity theorist. As an important example of his concern for speed, he gave a great deal of thought to minimizing the time that carries might add to the time for the addition operation. For example, adding 1 to 999...9 would, in the most straightforward schemes, require propagating a carry to the left for every one of the 50 digits that Babbage intended to use. He finally arrived at a method that he called the "anticipating carriage," avoiding this falling-domino effect (called *ripple carry*) and beating the patent for the modern carry-lookahead adder by more than 120 years.[10]

Collier (1970) mentions some other remarkably prescient refinements planned for the engine: precomputed tables in external memory; programmable formats for output; and automatic error detection, with the engine ringing "a great bell to tell [the] Assistant he had made a mistake." One wonders just how great a bell Babbage had in mind.

9.8 Augusta Ada Byron, Countess of Lovelace

Despite the fire in Babbage's brain, and his many breakthroughs, his work was being increasingly ignored in his own country. Hardly anyone understood what he was driving at, which, given the extraordinary leaps of his imagination, is understandable; and without a working model of a machine, he was increasingly ignored by his compatriots. He did, however, have a sympathetic ear in his Italian friend Giovanni Plana, who invited him to present his ideas at an 1840 meeting of distinguished Italian scientists in Turin.

The Turin meeting was a success for Babbage, and the general reaction from the top scientists of Italy was enthusiastic. Most important was the fact that Luigi Menabrea heard his presentation, because Menabrea published a report of Babbage's analytical engine in a Swiss journal in 1842.[11] Menabrea's paper was written in French—and were it not for that fact we would know much less today of a certain Ada Lovelace, or perhaps even of Charles Babbage himself.

Ada Lovelace was the daughter of Lord Byron and a brilliant, mathematically gifted, and confident young woman—a difficult circumstance in Victorian England, on all counts. She met Babbage when she was a mere seventeen, and, shortly after, when she saw a demonstration of the working piece of his difference engine at one of his soirées, she was captivated and remained his lifelong friend and supporter. The full story of Ada Lovelace and her relationship with Babbage and his engines is both inspiring and, in some ways, poignant. But it would distract us from our path to the construction of an ideal discrete machine, so we remain focused on her key contribution.[12]

The well-known scientist and inventor Charles Wheatstone proposed to Ada Lovelace that she translate Menabrea's paper, and this she did, in splendid style, with the addition of extensive "Notes" (as she called them), longer than the original paper itself.[13] Her note G is famous for being, perhaps, the first computer program; it calculates what are called Bernoulli numbers.[14]

Lovelace collaborated closely with Babbage on her notes to Menabrea's paper, and just how much of the mathematical and programming content is original with her and how much was due to Babbage we leave as grist for the historian's mill. But her notes for the Menabrea translation are the fullest and clearest exposition we have today of Babbage's ideas for his analytical engine, and we are at the least deeply indebted to Ada Lovelace for her work in bringing them to publication.

9.9 Turing's Abstraction

There, in 1843, you have the modern computer: the stored program of Jacquard and the conditional execution of Babbage. Almost a century would pass before Alan Turing formulated an appropriate abstraction for it. Technology subsequently caught up with the idealization during the war years, and by the late 1940s, the extraordinary blossoming of the digital computer was well under way.

Looking back on a very eventful century and a half, it seems simple, with the lucidity of hindsight, to put the pieces together. Turing wanted to investigate questions about what a "machine" could do. Put yourself in his place: How would you go about it? How would you construct an ideal machine that is, on the one hand, as simple as possible, and on the other, does everything important that a digital computer does?

Consider first the input data. Given our earlier discussion of the virtues of keeping information, not only in discrete but in binary form, let's stick with just 0s and 1s. There is no simpler way to arrange the binary data than in a straight line, so let's use a *tape*, visualized as being divided into cells, each of which contains either a 0 or 1. How long a tape? We're free to make the rules here,

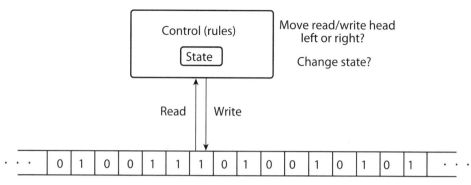

FIGURE 9.2. The Turing machine abstraction.

so let's make the tape extend as long as we want, to the left and the right.[15]

If you peek ahead at the picture of the completed, hypothetical machine in figure 9.2, you can see the doubly infinite tape at the bottom. We're committed to separate the program from the rest of the machine, and the figure shows it, labeled "control," in its own box. The control must be able to access the data, meaning read and change it, and this is shown as arrows between the control (program) box and the data on the tape. According to plan, we choose the simplest possible arrangement for this: At any given time, the control reads the contents (0 or 1) of a particular cell on the tape and can change the contents only of that cell. In analogy with an old-fashioned tape recorder or a more modern magnetic disk, we imagine a read/write *head* that is positioned at a particular cell position on the tape.

What's left is to decide how the control unit works. We don't have much choice: it will be a finite list of rules, each rule specifying what happens when the head reads a 0 and what happens when it reads a 1: Do we change the bit scanned by the head, and then do we move the head to the left or right?

At this point we seem to get to a roadblock. The rules look like this: "If the head is reading a 0, either change it to a 1 or leave it alone; then move the head either left or right on the tape." There are only two rules in the control, which depend on whether the head is reading a 0 or 1, and when you take into account the symmetrical variations, there just aren't very many possibilities.

Some rule-pairs always move in one direction, some leave the tape unchanged, but none of them leads to very interesting possibilities.

We need a new idea, and the impotence of the model so far proposed can be traced precisely to Babbage's insight of 1843: the results of computations succeeding from a given point should depend on the results of previous computations. While it is true that some of the bits on the tape of our model get changed and may affect future steps, the dependence is too limited to be of any value in making a powerful machine. There is more than one way to fix things, and Turing chose to use the idea of a *state*.

We assume, along with Turing, that the machine is, at any given time, in one of a finite number of predefined *states*.[16] The rules, then, depend not only on whether the head is scanning a 0 or 1 but also on the current state of the machine. The rules then stipulate not only whether or not to change the scanned bit and whether to move left or right, but also what the state will become after the current step is completed. Figure 9.2 shows the state of the machine stored in the control box.

The addition of a state has the wondrous effect of making the Turing machine as powerful, in theory, as any computer we know how to build. More about that in the next chapter.

I mentioned that adding state is not the only way to make a model of a computing machine that is equal in power to today's machines or, for that matter, any that we can *imagine* constructing. The importance of the state is that it allows the evolution of the machine to depend on past computations. We can also accomplish this, alternatively, by allowing the head to scan more than one cell at a time. The resulting machines are called *cellular automata*, and with a head scanning as few as three cells at once, say, and the appropriate rules, it is known that they can do any computation that a Turing machine can.[17] The study of cellular automata goes back at least to von Neumann, who was studying self-replication.[18] We have neither the need nor the space to discuss them here, but for us the most important fact about cellular automata is precisely that the interesting ones have the same computational power as Turing machines—they are the result of taking a different turn on the road to construction taken in this chapter, but the road leads essentially to the same place. Sometimes I like to think of them as "crypto Turing machines," but we might just as well think of Turing machines as "crypto cellular automata."

This gets to the main point of our discussion of Jacquard's stored program and Babbage's conditional execution: there is a notion of *computational power* that is captured by the Turing machine, and embraces all digital computers and—perhaps—all computers. Consider this statement a challenge and an invitation to the subject of computing to solve problems, and the next two chapters.

10 Intrinsic Difficulty

10.1 Being Robust

We now focus on the question of how fast computers can solve various kinds of problems. Many theoretical computer scientists devote their research to this field, called *computational complexity*.[1] The initial challenge in studying these sorts of questions is to boil things down to essentials and not get bogged down in details. There are many different kinds of digital computers, many different languages that are used to code algorithms to solve problems, and many different algorithms that can be used to solve any given problem. How do we go about formulating questions that have general significance? Getting to the heart of the matter took 35 years, from 1936 to 1971.

I mark the incubation period of complexity theory as precisely as I do by the dates of two truly remarkable and influential papers: Turing (1936) and Cook (1971). As mentioned above, the challenge in attacking general questions about the difficulty of problems, and many other scientific questions, is to get results that are broadly applicable, avoiding a morass of details. For example, we don't want to spend a great deal of time proving that a certain problem can be solved fast on Windows machines but leave the question open for Macs. By the same token, we don't want to spend a lot of time proving something about finding a schedule for delivery trucks that tells us nothing at all about scheduling steamships. What we need is a way of defining computing machines that is *robust* enough to include any reasonable type of machine we might use, and a way of looking at problems that is robust enough to distinguish between easy and hard problems in a practical sense. In his 1936 paper, Alan Turing defined the computer in a precise and fruitful way; and in his 1971 paper, Stephen Cook gave us a working definition of an "easy" problem and a way of looking at the larger picture that provided the all-important abstraction from detail.[2]

10.2 The Polynomial/Exponential Dichotomy

The framework for studying how long it takes computers to solve problems has become quite standardized. We assume that every instance of a problem comes with a number, called its *size*, that tells us roughly how "big" the given instance is. To be more precise, the size of a problem instance is the number of symbols it takes to write down the data that specifies its input. We then study how many steps a given algorithm takes to solve an instance of a problem, as a function of the size of the instance. Usually, we worry about running out of computation time for large instances, so we can ignore the running times of the algorithms on small examples, focusing attention on what happens when the size of the instances gets larger and larger. The resulting time requirement is called the *time complexity* of the algorithm.[3]

For example, suppose we want to sort the names on a list, putting them in alphabetical order. An instance of this *sorting* problem is a list of names, and we can take the size of an instance to be, simply, the number of names on the list. As a second example, if we want to study algorithms for solving the Steiner problem that we discussed in chapter 8, we can regard the number of cities as the size of an instance. In both examples we make the common and reasonable assumption that each data item, such as a name or city location, fits in a fixed number of storage locations. This justifies our thinking of the size of an instance as the number of data items.

As you might know or imagine, sorting things is such a simple, fundamental, and commonly encountered problem that it serves as a traditional early object of attention in introductory computer science courses. The most naïve and straightforward algorithms for sorting take a number of steps proportional (for larger and larger instances) to the number of items squared. There are, however, better ways of doing things, and ways of sorting in time proportional to only the number of items times its *logarithm* provide first moral guidance to potential sinners in the world of algorithm design. The Steiner problem, however, is another matter entirely; even the best-known algorithms for it require exponential time.

The key to the robustness in both Turing's and Cook's work is the distinction we've seen before in connection with Moore's law. We can use it as a guiding principle: *A time complexity that grows as a polynomial is practical and acceptable; an exponential rate is not.* We've already noted that an exponential rate will run us, inevitably, into a brick wall. A polynomial rate is *qualitatively* gentler. When an algorithm runs in time that is a polynomial in the size of the problem, we say the algorithm is *polynomial-time*, and similarly for exponential.

Returning to the two problems we used as examples above, in professional jargon: There are many ways to sort in polynomial time, but no polynomial-time algorithm has yet been found for the Steiner problem. Furthermore, there is good reason to believe that no polynomial-time algorithm for the Steiner problem exists, and the evidence for this is the subject of the rest of this chapter. In a practical sense, sorting is easy and the Steiner problem is, as far as we know, hard.

The polynomial/exponential dichotomy we use to distinguish between "easy" and "hard" problems is actually a liberal license to be sloppy. For example, we haven't said what unit we should use to measure time. Should it be centuries, microseconds, or machine cycles? The difference is only a scale factor, and a polynomial stays a polynomial if we rescale the time by any constant. What about the kind of computer we use? Turing defined his machine very precisely because he wanted to prove theorems about it, but (fortunately) we don't run our programs on Turing machines. How can we draw conclusions about everyday computational requirements from theorems about Turing machines, which are highly idealized? Again, the polynomial/exponential dichotomy rescues us, because we can prove that the time requirement of an algorithm on any reasonable kind of computer is a polynomial in its time requirement on any other reasonable kind of computer. Because the dichotomy embodies the vast difference between polynomials and exponentials, these proofs do not usually require much delicacy.

10.3 Turing Equivalence

To go further in explaining the way we compare computing machines using the polynomial/exponential dichotomy, we need to discuss what it means for one machine to *simulate* another. It is a simple but important notion: one machine simulates another if they both produce the same output when run with the same input. One machine may be much faster than the other, and the internal workings may be completely different. We insist only that their input-to-output behavior be the same.[4]

To take an example, consider an ordinary desktop computer, which runs programs written in some language—it doesn't really matter which—and produces a definite output given a definite input. The desktop can certainly simulate a Turing machine. In fact, at any given time there are scores of such simulators available online. Turning it around, is there a Turing machine that will simulate the desktop? There is no need to go into details here, but with a little experience it is not really difficult, in principle, to design a Turing machine that will execute the machine language of a given desktop. It is not something that makes much sense to do except as an exercise, but there is no question that it can always be done.

We see, then, that a typical desktop and a Turing machine can simulate each other. When this is the case, we say the machines are *equivalent*, and if a machine is equivalent to a Turing machine, we say it is *Turing equivalent*. Thus, we have argued that an ordinary desktop is Turing equivalent. We have said nothing, however, about how efficient the simulations involved can be. It may be, for example, that a machine can be simulated with a Turing machine but only by using exponential time. More interesting to us is the situation when the simulations involved in machine equivalence are polynomial; that is, when each machine executes in time that is polynomial in the time used by the other. When this is true, we say the machines are *polynomially equivalent*. It turns out that a Turing machine can be programmed to simulate any ordinary desktop with only a polynomial penalty in running time, and vice versa. In this case we say that desktops are *polynomially Turing equivalent*. Your smartphone, laptop, and the chip in your dishwasher and car are all polynomially Turing equivalent. Any device that truly earns the adjective "digital" can simulate a Turing machine in polynomial time, and vice versa.

A key question arises here: Are *all* computers, including the analog computers discussed in the previous chapter, polynomially Turing equivalent? We postpone this question to the next chapter, restricting ourselves here to digital computers.

It may happen that simulating a desktop on a Turing machine is a rather inefficient operation, with the Turing machine requiring, say, the square of the number of steps that the desktop does for each problem. This can happen easily. For example, the Turing machine may have to run back and forth in its memory to simulate fetching a piece of data. At first this might seem disastrous. After all, squaring the computation time means 100 seconds becomes 10,000 seconds. But the square of a polynomial is still a polynomial, and for this and similar reasons we can always simulate a desktop with a Turing machine while preserving the polynomial nature of any algorithm's time complexity. What counts for us is staying on the polynomial side of the dichotomy.

At this point it is worth emphasizing once more the qualitative difference between polynomial and exponential growth rates. We can almost always regard the former as benign, but the latter is likely to evoke the term "brick wall." I used that term in connection with Moore's law, and I meant by it that the transistors on a chip are becoming so small that we must run into a hard physical limit in a very few years. A transistor cannot, as far as we know, be smaller than an electron. The exponential law of doubling the density of chips every two or three years (or ten for that matter), implies that we are headed for a collision with very solid, unyielding reality.

Now let's do a little arithmetic to see what an exponential growth rate might mean for the running time of a program. Suppose, for example, that we want to process a list of names and that it takes $N^2/100$ seconds to process N names. A list of 10 names takes 1 second. Processing a list 10 times longer, 100 names, then takes only 100 seconds, not enough time to brew a cup of tea while we wait.

Suppose instead that a second sorting program takes $2^N/1000$ seconds, just to make it comparable to the first program for a list of 10 names ($2^{10}/1000$ is about 1 second). Processing 100 names then takes about 40,196,936,841,331,475,187 years, which is enough time to brew an awful lot of tea. This illustrates just why the polynomial/exponential dichotomy is our great friend: the contrast between the two kinds of behavior is so great that we can ignore messy details in analyzing algorithms while still preserving the crucial distinction between the practical and the absurd.

10.4 Two Important Problems

We now consider two very well-known problems, ones that you will inevitably meet if you go on to study computational complexity in any depth. To explain exactly what these problems are, we need some notation that we've already used, without fanfare, in chapter 3, when we discussed building logic gates from valves. I'll set down the little we need in the next two paragraphs.[5] The reward for using this modicum of formalism will be an acquaintance with Cook's theorem, the central result of theoretical computer science.

Recall that when we discussed signal standardization in chapter 3, we dealt with discrete signals that can take on either the value TRUE or FALSE. Logic gates take values of input signals and produce output signals, also either TRUE or FALSE, and we discussed NOT gates, AND gates, and OR gates. The two problems we now define are specified by logical expressions, or formulas, using these ingredients: variables a, b, c,..., and so on, each taking the value TRUE or FALSE; and the basic gate operations NOT, AND, and OR. For example, a typical logical expression might look like $Q = (a \text{ OR } b) \text{ AND } (b \text{ OR } c)$. (This means that either a is TRUE or b is TRUE *and* either b is TRUE or c is TRUE.) We can also use NOT gates, and for convenience we put a bar over a variable, as in $\bar{x}$, to stand for NOT x. (If x is TRUE, $\bar{x}$ is FALSE, and vice versa.)

For stating our two problems, we want to consider logical expressions in a certain standard form, which is, briefly, the ANDing of ORs. That is, we restrict the formulas to look like the one we called Q above, and say that these logical formulas are in *conjunctive normal form* (CNF). When there are exactly two variables in every clause and the clauses are ANDed (as in Q above), we say a formula is in 2-CNF form. When there are exactly three variables in every clause and the clauses are ANDed, we say a formula is in 3-CNF form. For example, $R = (a$ OR b OR $c)$ AND $(\bar{b}$ OR c OR $\bar{d})$ is a 3-CNF formula.

We can now define our two problems, the first of which is famous and the second of which is infamous. Both problems are framed as yes/no questions, given an input formula in CNF form:

- **2-SATISFIABILITY (2-SAT):** Given any formula in 2-CNF form, can we choose TRUE/FALSE values for the variables a, $b, c, \ldots$ that make the formula TRUE?
- **3-SATISFIABILITY (3-SAT):** Given any formula in 3-CNF form, can we choose TRUE/FALSE values for the variables a, $b, c, \ldots$ that make the formula TRUE?

Although these two problems may seem almost the same, the similarity is quite deceptive. The first is like the sorting problem mentioned above, because there is an algorithm that solves it in an amount of time that is a polynomial in the length of the input formula. In the standard terminology, we say that 2-SAT can be solved in *polynomial time*. The second problem is a different kettle of fish, being like the Steiner problem. Many brilliant researchers have been trying to find a polynomial-time algorithm for 3-SAT for many years, with no success. On the other hand, and this should be kept in mind, no one has actually *proved* that there is no polynomial-time algorithm for 3-SAT. At this point it may seem that we still know absolutely nothing about what we term the *intrinsic difficulty* of problems. But the actual state of knowledge is both more subtle and more useful than that, thanks to some intriguing ideas, which we explore next.

10.5 Problems with Easily Checked Certificates (NP)

Notice that both of our two specimen problems, 2-SAT and 3-SAT, have a very convenient property. If I claim that a particular instance has a satisfying assignment (is a yes instance), then there is a way I could prove it to you quickly: I could produce such an assignment, and you could check it in polynomial time by simply substituting the assignment's TRUE/FALSE values for

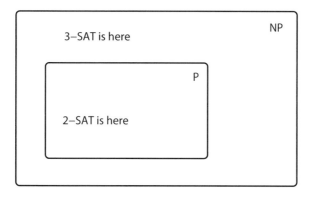

FIGURE 10.1. The class P is in NP. Are P and NP actually different? No one knows for sure.

the variables in the instance and verifying that each clause is rendered TRUE. This requires just one run through the instance, and therefore takes only polynomial time.[6] A list of symbols for a yes instance of a problem that can be checked in polynomial time is called a *certificate* or, sometimes, a *succinct certificate*. Of course, a certificate can be only polynomially long—otherwise it would take more than polynomial time merely to read it.

With this idea we can now define a very important class of problems, called NP, which are all those problems that have easily (polynomial-time) checkable certificates for every yes instance.[7] Both 2-SAT and 3-SAT are in NP because, as we've pointed out above, a list of TRUE/FALSE values for a yes instance can be verified quite easily. NP is a very large class and includes almost any yes/no problem you can think of, including both easy and hard ones.

We also define P (for *polynomial*) to be the very natural and important class of *easy* problems, those that can be solved in polynomial time. It is not hard to see that every problem in P is also in NP: if a problem is in P, just run the machine that answers the yes/no question. A record of that run can then serve as a certificate for the corresponding instance; it can be only polynomially long, since the machine that solves the problem in P executes only a polynomial number of steps. Figure 10.1 shows what is definitely known to be true: P is in NP, both 2-SAT and 3-SAT are in NP, and 2-SAT is in P.

10.6 Reducing One Problem to Another

To take stock, we are after some understanding of why some problems seem to be intrinsically difficult while others are easy. Unfortunately, these secrets about the computational power of digital computers are not completely understood and, in fact, may never be. But what we do understand can be traced back to the contributions of Turing and Cook mentioned at the beginning of this chapter. The last piece of machinery we need is the wonderfully fruitful idea of reducing one problem to another, which was exploited in Cook (1971).

To take a concrete example of a reduction, we consider two more very famous and important problems, the traveling salesman problem (TSP) and the Hamilton circuit problem (HC). In the general TSP, we are asked to choose the most efficient itinerary for a hypothetical traveling salesman. More precisely, we are given the distance between every pair of a given set of cities, and we are asked to find the tour, starting at the salesman's home city, visiting every other city *exactly once,* and then returning home, with minimum possible total distance. The mathematical problem has frustrated mathematicians and computer scientists for at least 80 years.

Besides its role as a celebrated mathematical puzzle, the TSP has considerable practical importance as well. For example, it comes up in choosing the order in which to drill holes in circuit boards: a computer-controlled drill on an assembly line must visit all the holes it is assigned to drill and return to its starting point for the next circuit board. The same problem comes up in scheduling the order in which a telescope visits a given set of locations in the sky. Problems with more than just a few cities quickly reveal the difficulty; there are an exponential number of possible tours, and it can easily happen that many of them are close in total distance. As a good illustration of just how hard this problem can be, even for reasonably small examples, Procter and Gamble ran a contest in 1962 with only 33 cities and offered a prize of $10,000 for the best solution.[8]

The second problem we consider, HC, is even older, and has been equally frustrating. In this problem, instead of being given distances between pairs of cities, we imagine a different situation: we are told that certain cities (not necessarily all pairs) are connected by direct roads, and we wish to answer the question of whether or not there is a tour that visits every city *exactly once* and returns to its starting point.

The TSP and HC seem to be closely related. In fact, they are related in the following precise sense: if we have an algorithm that solves TSP, then we can use that algorithm to solve HC. To see this, suppose we are given an instance of HC that we wish to solve. Then all we need to do is construct an instance of TSP where the pairs of cities that are connected in the HC problem are assigned the distance 0 and the pairs that are not connected are assigned the distance 1. Then if the minimum-distance tour for the resulting TSP has total length 0, we know that there is a tour for the HC problem since all the links used in the solution must all have length 0 and they correspond to roads that are present. If the minimum-distance tour for the TSP has total length greater than 0, then we know that there is no tour for the HC problem, for there is a missing road. In this situation, when we can solve HC using a program that solves TSP, we say that HC *reduces* to TSP. We always assume that the construction of the reduction is *efficient* in the sense that it takes polynomial time, as it does in this simple example.

Notice now that if problem A reduces to problem B, it means that if we know how to solve problem B in polynomial time, then we can also solve problem A in polynomial time. In other words, *if A reduces to B, then B is at least as hard as A.* We use reductions in this way, starting with one problem and creating chains of reductions to show that many other problems are at least as hard as the one we started with. This modus operandi can be a source of confusion the first time you see it, because there is a natural tendency to look for ways to reduce a problem to an *easy* one; solving the latter then solves the former—thus exploiting the fact that reducing problem A to B shows that A is at least as *easy* as problem B. Here we turn the logic around: We reduce A to B to show that B is at least as *hard* as A.

10.7 Yes/No Problems

I should mention a detail about the way problems are specified. For simplicity, I have been assuming that they are framed as yes/no questions, like 2-SAT, 3-SAT, and HC. The traveling sales-man problem, however, asks for a tour with minimum distance. This complication very often turns out to be unimportant in practice, because we can usually show that an algorithm that yields a yes/no answer can be used to construct a solution like a tour. For example, in the TSP, we can ask if there is a tour with a total length that does not exceed a certain number; then narrow the range by successive refinement; then ask the question when individual links are given an artificially high cost; and so on, ultimately finding a set of links in an optimal tour. What is important here is that we use the hypothetical yes/no algorithm only a polynomial number of times. Using a polynomial-time algorithm a polynomial number of times is another polynomial-time algorithm. Thus, in our example, an efficient algorithm for the yes/no version of the TSP (where we ask if a tour exists with a certain cost), would yield an efficient algorithm that actually produces a tour. Once again, the polynomial/exponential dichotomy greatly simplifies the work in studying algorithms and allows us to concentrate on the larger picture.

10.8 Cook's Theorem: 3-SAT Is NP-Complete

We are now in a position to understand Cook's theorem.[9] It is an amazing result, with far-reaching consequences: we show that every problem in NP reduces to 3-SAT.

The proof exploits the fact that 3-SAT can be viewed in two ways. First, we simply ask if we can find TRUE/FALSE values for the variables in a given 3-CNF expression that makes that expression TRUE. In this case we treat an instance of the 3-SAT problem as an abstract question and nothing more. But we can also assign *meanings* to the variables, and interpret the 3-CNF expression as a statement about things we care about. It is the latter interpretation that points the way to our goal.

Now, suppose we take an instance of any problem in NP, say, problem A. Because A is in NP, we know that there is a certificate for every yes instance of A that will be verified by a correctly functioning Turing machine. It turns out that from the instance of problem A we can construct an instance of 3-SAT that expresses precisely the following statement:[10] "There is a certificate for this instance of problem A that will be verified by a correctly functioning Turing machine." The TRUE/FALSE values for the variables in the expression correspond to the certificate, which is, after all, nothing more than a list of bits. The variables in the constructed certificate have interpretations like "at time 10 the 97th position of the Turing machine's memory is being used and the 42nd instruction is being executed." The 3-SAT expression will express things like "every location in the memory of the Turing machine contains one and only one allowable symbol," and so on.

Clearly, the expression we construct for any given instance will use many variables, and many clauses (which are all ANDed together), but when the dust clears—and this is crucial—it will be only *polynomially* long in terms of the instance of problem A that we started with. Furthermore, the constructed expression will have a satisfying choice of TRUE/FALSE values for its variables when, and only when, there is a certificate showing that the instance is a yes instance of problem A. This is exactly what we mean by claiming that A reduces to 3-SAT. Thus, any instance of problem A reduces to 3-SAT. If this were a mathematics book and the details of this proof were spelled out, we would be entitled to say "QED."

The immediate consequence of this result is that if we could solve 3-SAT efficiently, we could solve all problems in NP efficiently. This can be expressed by saying that 3-SAT is as hard as any problem in NP. A problem in NP that has this property is called *NP-complete,* and we can state Cook's theorem in a very concise way: 3-SAT is NP-complete. The idea of NP-completeness is very powerful, for the following reason. Suppose we could find a way to solve an NP-complete problem, say, problem X, efficiently (that is, in polynomial time). Then, because every problem in NP reduces to X, we would be able to solve any problem in NP efficiently.

This result is worth contemplating; it is the key to the most important idea in theoretical computer science. Restating it informally: Because any NP-complete problem is as hard as any problem in NP, cracking one NP-complete problem cracks them all. Restating it more formally: Showing that any problem in NP is in P shows that all of NP is in P—and therefore that P = NP. We can say it yet another way: finding a way to solve any one NP-complete problem efficiently would mean that if we can merely *check* a solution to a problem easily, we can also *find* a solution easily.

We are now in the same position as everyone else in 1971 when Cook presented his paper; we know only one NP-complete problem, 3-SAT. But that situation changed quickly and dramatically.

10.9 Thousands More NP-Complete Problems

The year after Cook published his theorem and introduced the idea of NP-completeness, Richard Karp showed, in a paper that can fairly be described as sensational, that many interesting, classic problems besides 3-SAT are also NP-complete.[11]

Karp used the following observation. Suppose we could reduce 3-SAT to another problem in NP, say, X. Then all of NP, which reduces to 3-SAT, would also reduce to X—a polynomial reduction of a polynomial reduction is a polynomial reduction. Therefore, X would also be NP-complete. This means that we can start with one problem known to be NP-complete, say, 3-SAT; reduce 3-SAT to X (showing that X is NP-complete); then reduce X to Y (showing that Y is NP-complete); and so on, producing a chain of NP-complete problems. Intuitively, if X is at least as hard as 3-SAT, and Y is at least as hard as X, then Y must be at least as hard as 3-SAT, and therefore all of NP. We can also branch out by reducing any one of these problems to two or more other NP-complete problems, producing what is called a *tree* of NP-complete problems. Cook's theorem provides the seed for this process; we would not be able to carry out this plan without a place to start (see figure 10.2).

In his 1972 paper, Karp executed exactly this plan, start-ing with 3-SAT (actually, a general version of 3-SAT), and by a series of reductions showed that 21 well-known, and apparently intractable, problems are NP-complete.[12] Figure 10.3 shows his reduction tree. As you can see, he showed to be NP-complete the Steiner problem that we described in connection with soap films in the previous chapter, as well as versions of the Hamilton circuit

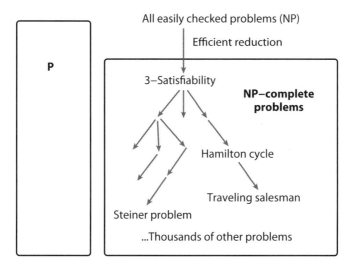

FIGURE 10.2. How NP-complete problems are defined, and how NP-completeness is proved. The reduction of any problem in NP to 3-SAT, which is Cook's theorem, appears at the top. A tree of reductions then produces the thousands of problems now known to be NP-complete. The "easy" problems are in P, and if *any* NP-complete problem is in P, they *all* are, P = NP. To repeat, no one knows for sure.

problem. The floodgates were now open, and a great variety of problems were soon shown to be NP-complete. By 1979 there were hundreds, collected in Garey and Johnson (1979). Today there are thousands. It is rare in any field that such a beautiful unification takes place, and it all depends critically on the polynomial/exponential dichotomy.

It is worth emphasizing the power of the idea: If *any one* of these NP-complete problems can be solved in polynomial time, then so can *every* problem in NP—which means almost any problem of a discrete nature you can think of. Solving any one of these thousands of problems would mean that P = NP (and instant fame). The fact that many very brilliant researchers have been trying to solve some of these NP-complete problems for a long time is very strong evidence that P ≠ NP and that NP-complete problems are, in fact, intrinsically difficult. The theory brings us as close as we may ever get to understanding why some problems are so much harder than others.

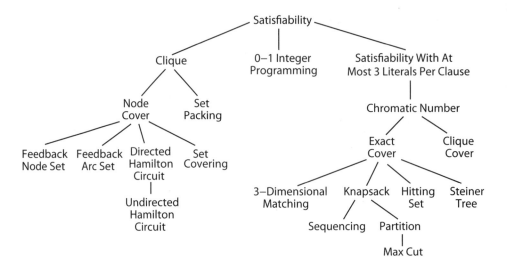

FIGURE 10.3. Karp's 1972 reduction tree. Notice our old friend Steiner Tree, as well as versions of the Hamilton circuit problem. (After Karp (1972), fig. 1.)

FIGURE 10.4. Bricks in the wall: The most important open question in computer science, encoded in the west wall of the Computer Science Building, Princeton University. Can you decode it? (For a spoiler, see https://www.cs. princeton.edu/general/bricks. Accessed September 15, 2017. Photo by the author.)

The glaring missing piece in all this, of course, is that so far, no one has been able to actually *prove* that P $\neq$ NP, even though almost every computer scientist believes it to be true. Repeated attempts to find a proof have failed, but so have attempts to find an efficient solution for an NP-complete problem. Every computer science theorist receives a regular stream of unsolicited purported proofs one way or another, and, so far, all the proofs have been shown to be faulty, with mistakes of varying degrees of subtlety. The P $=$ NP question is so fundamental and so puzzling that it is quite literally built into the building that houses your author's home department, as shown in figure 10.4.

Up to now, I have tried to lay out the reasons why the word *computation* today almost always implies *digital computation*. Our world has certainly been captured by digital technology, and when we speak of the limits of computation, it usually means the limits of computation by digital computer. But I planted some hints, in section 10.3 and in chapter 8, that there may be more computational power left hidden, somehow, in the analog world, and we devote the next chapter to that possibility.

11 Searching for Magic

11.1 Analog Attacks on NP-Complete Problems

A digital computer in the idealized world of theoretical computer science has only two resources: time and space (storage). Running out of time on large and difficult problems is usually the over-riding concern. From the theoretical point of view, the simplest model of a digital computer, the Turing machine, has a single read/write head that can move only one location at a time on the storage medium (the tape), and therefore the storage used cannot exceed the time used. As far as today's real computers are concerned, a half century of feverish hardware development has made memory dirt cheap. For these reasons, when we discuss digital computers, including Turing machines, we worry most about running out of computation *time.*

In the case of an analog computer, many other things can go wrong. For example, a machine may use an exponential amount of energy, its mass or size may grow exponentially fast, some parts may fail because of excessive stress, an insulator may break down because of excessive voltage, and so on. In evaluating an analog computer, we must therefore consider not only the time for a computation but also the possibility that its use of *any* resource becomes impractically large as the size of problems grows: time, space, energy, mass, material strength, and so on. Bear this in mind as we inspect a few particular analog machines.

Soap films for the Steiner problem

Revisiting the Steiner problem and its solution using soap films, we recall another complication that can doom an analog computer. As we saw in chapter 8, it seems at first that we can solve the NP-complete Steiner problem by simply dipping a wire frame into a soap solution. But something about the problem bites back. The fact that many possible configurations can be *local* but not necessarily *global* solutions wrecks the idea. Somehow the problem is what we termed "intrinsically difficult," but just how the difficulty manages to manifest itself in two seemingly disparate ways—having many local solutions on the one hand and being NP-complete on the other—is quite mysterious. You might get the feeling that God is trying to tell us that we are never going to find an efficient solution to the Steiner problem—nor any other NP-complete problem—no matter what kind of clever contraption we devise. This is, perhaps, another way of saying that there is a fundamental physical law at work, and we return to that thought below.

An electronic PARTITION machine

For another example of this phenomenon, we consider another NP-complete problem, called PARTITION.[1] As usual, the statement of the problem is deceptively simple: you are given a list of positive integers that sum to some number N, and you must decide whether it is possible to split the set into two subsets, each of which sums to $N/2$. For example, if you are given the set $\{8, 3, 16, 9, 21, 12, 3\}$, the answer is yes.[2]

The proposal for an analog machine to solve PARTITION is based on the fact that multiplying signals together produces new frequencies that are all possible sums and differences of the frequencies in the original signals.[3] A little more terminology allows us to be more precise. We use the term *sinusoid* to mean any single sine or cosine wave at a single frequency, a pure tone. The sum or difference of sinusoids of the same frequency is also a sinusoid of that frequency, even if one is shifted with respect to the others. If we now multiply together two sinusoids of frequencies f_1 and f_2, we get a signal that contains frequencies $f_1 + f_2$ and $f_1 - f_2$. This process is called *mixing* or *heterodyning*, and is used in almost every radio and television receiver to shift signal frequencies so stations can be more easily and accurately tuned in. Ham radio enthusiasts understand this technique very well and make good use of it in their transmitting and receiving equipment.

Continuing with more signals, multiplying three sinusoids together produces frequencies $f_1+f_2+f_3$, $f_1+f_2-f_3$, $f_1-f_2+f_3$, and $f_1-f_2-f_3$. Notice what is happening here: every time we multiply a sinusoid by an additional sinusoid, we double the number of frequencies in the signal, and at any point we have all possible sums and differences of the frequencies we are using. If we do this with frequencies that are the integers for an instance of the PARTITION problem, then among all the resulting frequencies we get the frequency *zero* when, and only when, we can add and subtract the frequencies in such a way that the positive and the negative contributions cancel—situations that correspond exactly to yes instances of PARTITION.

As discussed in chapter 8, the traditional general-purpose electronic analog computer is based on the operation of integration, or averaging. Such machines are now practically obsolete, but if we built one, we could use it to solve the PARTITION problem. First, it is an easy matter to generate a sinusoid at a given frequency, by feeding back the output of two stages of integration to its input, a very standard configuration that is used to generate sinusoidal signals. This works because of the mathematical fact that if you integrate a sinusoid twice, you get back to a multiple of itself—and the feedback loop enforces this relationship. Second, it is easy to generate sinusoids with the frequencies we need—f_1, f_2, f_3, and so on—by multiplying the original sinusoid by itself. We can do this with a special analog multiplier, but it is also possible to perform multiplication with integrators.[4] We can therefore multiply together all the sinusoids with frequencies that correspond to the input data for a PARTITION problem, producing an output waveform we call S or, since it is a waveform that varies with time, $S(t)$.

As argued above, the answer to the particular PARTITION instance we are dealing with is yes if the signal $S(t)$ contains the frequency zero, and no if not. The sinusoid of frequency zero is special: it is a *constant,* while sinusoids at all other frequencies oscillate. The average of a zero-frequency sinusoid will therefore always be some constant, not zero. If, however, we average a

Input data

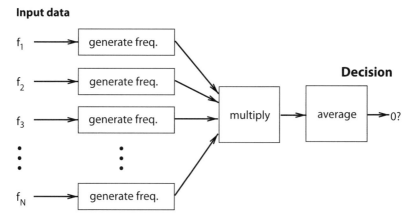

FIGURE 11.1. An analog machine for solving PARTITION. The input data enters from the left, and corresponding frequencies are then generated and multiplied together. The result is averaged, and the answer to the PARTITION problem is no if the result is zero, and yes otherwise. Why does this not work in practice?

sinusoid of any other frequency for one of its periods, or any number of its periods, the result will be zero, because the upswings and downswings will cancel. We can therefore decide whether the frequency zero is present in $S(t)$, and hence whether the instance of PARTITION is yes, by averaging it over some number of periods of its lowest frequency and seeing whether its average value is not, or is, zero. A sketch of the machine is shown in figure 11.1.

Does this analog machine solve the NP-complete problem PARTITION without something going haywire? Actually, it isn't very difficult to see intuitively what goes wrong. A hint is provided by the statement above that when we multiply the generated frequencies together, we get "all possible sums and differences of the frequencies we are using." There are, therefore, an *exponential* number of frequencies present, and they can add to produce an exponentially large signal, while the product of sinusoids is never larger than one. In fact, the mathematics shows that we need to divide by two every time we multiply by a sinusoid, and we end up trying to distinguish between zero and something exponentially small. This requires integrating the output signal an exponentially long time, because of the inevitable presence of noise in the world. The machine is thus stymied by two old enemies: exponential growth and noise.

A gear machine for 3-SAT

I like to think of the third example of an analog attack on an NP-complete problem as a direct descendant of the Antikythera mechanism, because of the contribution by Anastasios Vergis.[5] It is meant to solve 3-SAT, and consists of a gearing mechanism with the addition of smooth cams and limit stops that prevent shafts from rotating past certain positions. The machine requires only a polynomial amount of material to build. The information about the clauses in the 3-SAT instance is encoded into the construction of the machine through the gear linkages, and there is one particular shaft with a crank handle. If you push on the crank handle and it moves, the answer to the original 3-SAT instance is yes; if it doesn't, the answer is no.

The details of how to go from a particular 3-SAT problem to such a gear machine are a bit involved, and we skip them here. But I have given you abundant reason to be very suspicious without even looking at how the machine is supposed to work. Likely something goes wrong. Why can't a mechanical device like this solve the NP-complete problem 3-SAT using polynomial resources? Exactly what does go wrong? Frank Lee devoted a master's thesis to analyzing the 3-SAT machine,[6] and Michael Main designed a traditional electronic analog computer version of the machine.[7] I think it's fair to say that neither Lee nor Main offers a decisive annihilation, although their analyses are quite cogent and they cite several very reasonable misgivings about the purportedly efficient operation of the machine. It is, in fact, the multiplicity of the objections that gives me pause. The authors of Vergis et al. (1986) discreetly left the question unanswered. Thirty years later, I, for one, still do not know exactly what is wrong with the gear machine for 3-SAT, but, for the reasons given above, I am now even more certain that there *is* something fundamentally wrong with it.

Many other analog machines have been proposed for solving NP-complete problems, and I again recommend Scott Aaronson's review,[8] at least for the technically more advanced reader. The interested reader may also find it entertaining sport to collect other proposals for NP-complete machines and debunk them.

Oltean (2008), for example, proposes a machine for solving the Hamilton path problem (a slight variant of the Hamilton circuit problem, also NP-complete), which is based on the propagation of light through optical cables. A much more complex and challenging example is the machine for solving SUBSET SUM (a variant of PARTITION, also NP-complete) described in Traversa et al. (2015). Judging from the title of the paper, "Memcomputing NP-complete problems in polynomial time using polynomial resources and collective states," the authors apparently believe the machine will work in the sense of scaling up to larger and larger problems; by now you know where I put my money.[9]

11.2 The Missing Law

We've seen analog machines based on widely different physical principles all fail to solve NP-complete problems; what is remarkable is that they fail for apparently very different reasons. The soap-film computer for the Steiner problem fails because of the unmanageable multitude of local minima, and the integrator-based machine for PARTITION fails because of noise. It appears that the gear machine for 3-SAT fails because of some mechanical difficulty associated with the limited accuracy of machining or the transmission of forces through gear trains. The persistence of practical difficulties of different kinds in such disparate implementations suggests that there is a fundamental physical law at work. Lee (1999) points out that the situation is analogous to the perennial proposals for perpetual motion machines, which are today rejected out of hand because they would violate the first or second law of thermodynamics. If some day the US Patent Office automatically rejects proposals for analog machines to solve NP-complete problems, what physical law will be invoked in the form letter of rejection?

Note that the fundamental law we are looking for cannot be $P \neq NP$, for that is, after all, a mathematical conjecture, and can say nothing directly about the physical world. To pinpoint the law, we turn once again to our physicist-in-residence, Richard Feynman, as well as Alan Turing and, as it happens, Turing's thesis adviser, Alonzo Church—certainly a celebrated consulting committee.

11.3 The Church-Turing Thesis

Turing (1936) invented a simple and concrete description of a hypothetical machine, which, naturally, we now call the *Turing machine,* in order to study a fundamental question about the nature of computation. He was interested in what numbers could be written down by a machine. To review: A Turing machine has a memory in the form of a tape (as long as it needs to be); a head that reads and writes symbols (from a finite alphabet) from and to the tape; and a stored, fixed program that, consulting the machine's state, controls what the head does. He very successfully captured the notion of step-by-step computation, and with the license of polynomial equivalence, it is today still the iconic digital computer for theorists. At about the same time, Alonzo Church published what turned out to be an equivalent definition of a computer, using a system for manipulating symbols called the *lambda calculus.*

The interrelationships among the contributions of Turing and Church, as well as Kurt Gödel, Stephen Kleene, Emil Post, J. Barkley Rosser, and others, are complex, and best left to historians of science. Fortunately, we need only the one simple (in retrospect!) idea that there is such a thing as a "computing machine." Turing (1936) envisioned the "computer" as a human or automaton following a definite, finite set of instructions, writing symbols down with a pencil. In the first paragraph of that paper he states, "According to my definition, a number is computable if its decimal can be written down by a machine." What is very interesting, and perhaps surprising, is that there are numbers that cannot be so written down. The proof can be easily, if roughly, paraphrased in one sentence: it is possible to count, one after another, all the Turing machines (each programmed to write down one number); but it is not possible to count all the possible numbers.[10]

The *Church-Turing thesis* emerged from a burst of fundamental work in logic in the 1930s. It is not a mathematical statement, which might be proved or disproved, but is, rather, a premise, a hypothesis, that is informally expressed. It is not provable, because it is a statement about physics.[11] The thesis uses the concept of Turing equivalence given in the preceding chapter, and states: *Any reasonable computer is Turing equivalent.* That is, any reasonable computer can simulate a Turing machine, and a Turing machine can simulate any reasonable computer. Yao (2003) puts it this way: "The Church-Turing Thesis (CT) is the belief that, in the standard Turing machine model, one has found the most general concept for computability."

We can certainly regard analog computers as "reasonable," and this returns us, once more, to our central theme—the relative virtues of analog and digital machines. At first, we might worry about the fact that a Turing machine is digital in nature, having only a finite number of allowed symbols to manipulate, while an analog computer can involve continuous quantities. But that isn't really a problem for gauging computational power; we can directly compare any analog computer to a Turing machine if we regard the analog computer as a device that yields only a yes/no answer to any given problem. The Church-Turing thesis goes much deeper than that, and has provided employment for a small army of philosophers over the years.[12]

11.4 The Extended Church-Turing Thesis

The Church-Turing thesis posits the existence of a Turing machine that can simulate any computer, and vice versa, but says nothing about *efficiency*. In the 1970s the theory of computation was sharpened to take into account the polynomial/exponential dichotomy, and the correspondingly sharpened version of the Church-Turing thesis is now called the *extended Church-Turing thesis*.[13] Its precise origin is, to my knowledge, rather cloudy, but Richard Feynman, who was known for working things out for himself, stated a version of it in Feynman (1982):

> The rule of simulation that I would like to have is that the number of computer elements required to simulate a large physical system is only to be proportional to the space-time volume of

the physical system. I don't want to have an explosion. ...If doubling the volume of space and time means I'll need an exponentially larger computer, I consider that against the rules.

As suggested in Vergis et al. (1986), Feynman likely meant (or should have meant) "polynomial in the space-time volume" instead of "proportional to." The statement of the extended Church-Turing thesis then becomes, using the terminology of chapter 10: *Any reasonable computer is polynomially Turing equivalent.*

The extended Church-Turing thesis, by restricting the simulation time to be polynomial, allows us to say something very interesting about NP-complete problems. Suppose we could build an analog computer that actually solved an NP-complete problem in polynomial time. Then, by the extended Church-Turing thesis, that analog machine can be simulated *in polynomial time* by a Turing machine, and that Turing machine would then solve an NP-complete problem (and hence all NP-complete problems) in polynomial time. That would imply that $P = NP$, which, at this point, is something that most computer scientists believe to be false. Something has to give in this chain of reasoning, and the consensus is that, given the very strong evidence that $P \neq NP$, plus the strong, but perhaps not quite so strong, evidence for the extended Church-Turing thesis, there is no analog computer that can solve an NP-complete problem efficiently. It would appear that we have reached an impasse; trying to solve NP-complete problems efficiently with analog machines would seem to be futile.

We have not, however, reached an absolute dead end. A new avenue to explore was opened up by the same Feynman paper quoted above.

11.5 Locality: From Einstein to Bell

Recall that at the beginning of chapter 8 we excluded quantum mechanics from the discussion of analog computers. If we now allow quantum-mechanical machines, all bets are off. We now outline, broadly, the argument in Feynman (1982), a paper that has become famous as the spark that ignited the field of quantum computing. It was originally a conference keynote address and is very readable with just a little background.

Feynman addresses the question of whether there is a classical, nonquantum computer that can simulate a particular quantum-mechanical experiment. He makes some demands on the properties of the computer, two of which are important to us. First, he equips his computer with the power to make random decisions, a feature we have not yet considered. Quantum mechanics, which is what he is trying to simulate, is inherently probabilistic. That is, in general, the outcome of an experiment is not determined beforehand but is chosen randomly from a set of possible outcomes, with probabilities given by the theory. This is a feature of quantum mechanics that reportedly made Albert Einstein very uncomfortable, and he is often quoted as saying, "God doesn't play dice with the world." Feynman turns the tables in thinking about simulating quantum mechanics with a computer, and *demands* that his simulating computer be allowed to flip coins. The probabilistic version of a Turing machine has actually become the accepted model of what computer scientists regard as the standard (nonquantum) computer, the real and practical computer in your pocket or on your desk. There is nothing suspicious about a probabilistic Turing machine, and it is considered perfectly acceptable to allow it in our statement of the extended Church-Turing thesis. We might build the randomness feature into a real computer with a *pseudorandom* source, which would be generated by some very complicated and unpredictable program; or, for the purist, with a genuinely random source, which would ultimately be derived, naturally, from a quantum-mechanical process, like the radioactive decay of atomic nuclei.[14] Let's allow the hypothetical computer to flip coins.

The second important demand that Feynman makes for his computer is at the heart of the matter. He insists that his computer be *locally interconnected.* In his words, "I would not like to think of a very enormous computer with arbitrary interconnections throughout the entire thing." By this he means that, in simulating physics at a point, the computer can use only information that is available near that point. Note that this is *not* a

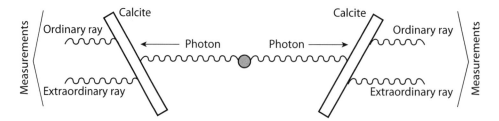

FIGURE 11.2. The hypothetical experiment used to show that quantum mechanics can violate Bell's inequality, and hence that quantum mechanics cannot be simulated by a computer that uses only local information. The atom in the center emits two entangled photons, and polarization measurements are taken at widely separated locations at the left and right. (After Feynman (1982), fig. 4.)

restriction that we can apply to a Turing machine, which can store any information we want. No matter. Feynman goes his own way and uses this picture to represent what is to him a reasonable computer, and it leads to a most stimulating proposal.

Feynman's locally connected computer fails to simulate quantum mechanics, and his conclusion is actually a version of a celebrated result called *Bell's theorem*, after J. S. Bell (1964). The theorem establishes an inequality, called *Bell's inequality,* that must hold for any computation that uses only locally available information. A hypothetical experiment then shows that quantum mechanics violates Bell's inequality, which therefore proves that a locally connected computer cannot simulate quantum mechanics.[15]

The particular thought experiment used by Feynman uses photons and calcite crystals, but there are many other systems that work equally well.[16] We begin with an atom that emits two photons, simultaneously, in opposite directions, as shown in figure 11.2. This can happen, for example, when a hydrogen atom loses energy. Photons have a polarization, which can be visualized as an arrow rotating in the plane perpendicular to the direction of travel. This picture should not be taken too literally, but it does have a basis in electromagnetic theory because the photon can be considered as a wave with rotating electric and magnetic fields at right angles to the direction in which the photon travels. We then know that the two photons must be spinning in opposite directions by a fundamental law of physics, the conservation of angular momentum.

In quantum mechanics the two photons cannot be considered separately, and the pair together is called an *EPR pair*, after a very famous paper by Einstein, Podolsky, and Rosen.[17] Bell (1964) was, in fact, a response to the objections raised in Einstein et al. (1935) to the apparent nonlocality of quantum mechanics. The two photons are said to be *entangled*, because of the complicated relationship between them. Without going into details, Bell's inequality is derived by measuring the polarization of the two photons after they have been widely separated, using calcite crystals, as shown in figure 11.2. The photons must be far enough apart so that no information about one photon can travel to the other without violating nature's speed limit, the speed of light. This is the origin of Feynman's locally connected condition. In Feynman's version, Bell's inequality says that the result of a particular set of measurements can never be larger than 2/3. On the other hand, quantum mechanics predicts the result to be 3/4. As Feynman puts it, "That's all. That's the difficulty. That's why quantum mechanics can't seem to be imitable by a local classical computer."

As mentioned above, the question of *locality* is central in this line of reasoning. That question, in fact, stimulated a considerable amount of further work on whether there are "hidden variables" that can and should be added to quantum mechanics to complete what some consider to be an incomplete description of "reality."[18] It was the nonlocality of quantum mechanics that bothered Einstein, Podolsky, and Rosen, and was shown to be unavoidable by Bell. Despite the anxiety at high levels about the nonlocality of quantum mechanics, however, no one has yet found a logical contradiction in its laws—although, somehow, they do appear to come about as close to a contradiction as possible. And, of course, so far, they work exceedingly well.

It is here that Feynman makes a suggestion that has had far-reaching consequences: "nature isn't classical, dammit, and if you want to make a simulation of nature, you'd better make it quantum mechanical." We now call any computer that makes use of quantum mechanics a *quantum computer*.

11.6 Behind the Quantum Curtain

With the help of quite a collection of geniuses, we have arrived at a very interesting question: Are there useful things we can compute efficiently with a quantum computer that are beyond the reach of an everyday, classical machine? A few more geniuses answered the question in the affirmative.

The first step was showing that there is a particular computing task, however simple, for which a quantum computer offers a definite speedup over any classical computer. Deutsch and Jozsa (1992) did exactly this, and the result, although seemingly trivial, strongly suggests that quantum computation might be the key to solving some critically important problems. Preskill (1998) describes an extremely simple case of the already extremely simple problem of Deutsch and Jozsa, and we present that next.

Suppose we are given a black box, called X, say, that has one input and one output; each input and each output is either 0 or 1. We are not allowed to look inside X and we know nothing about how it works.[19] Suppose also that X takes a long time to finish doing whatever it does, say, a year.

Now we are asked to determine if the output of X when its input is 0 is the same or different from its output when its input is 1. In the world of classical physics, the world of ordinary experience, the only way to decide whether or not the two outputs are equal is to first apply 0 as input, wait a year, apply 1, wait a year, and then compare the two results. We are assuming here that we have only one black box, so we cannot run two X's in parallel. This takes two years to get our answer, and it seems that there is no way around it.

It may surprise you, as it surprised me, that a quantum computer can answer Deutsch's question in one year instead of two. The method depends on the basic structure of quantum mechanics and the nature of quantum-mechanical measurement, and I will try to supply some intuition of the trick without doing excessive violence to the truth.

Quantum mechanics operates rather secretively, behind a curtain, you might say. In more mathematical terms, it operates in an abstract space, one that is not accessible to us unless we perform measurements. To take a simple concrete example, a photon can be put in a binary (two-valued) state, just as a valve or transistor can. These can correspond, for example, to the *polarization* of the photon—the direction in which its electromagnetic wave is rotating. They are traditionally called $|0\rangle$ and $|1\rangle$.[20] These states differ from the ordinary, classical states 0 and 1 in many ways, the most important being that a photon need not be in one state or the other but can be in both states at once, part $|0\rangle$ and part $|1\rangle$. Such a state is called a *superposition* state. In 1935 Erwin Schrödinger, to challenge the interpretation of quantum mechanics, devised a thought experiment in which a cat was put in a superposition of the states $|$alive$\rangle$ and $|$dead$\rangle$. Schrödinger's cat, as she is now called, has remained neither alive nor dead ever since.

Superposition takes place in an abstract space behind the quantum curtain, removed from our everyday experience. It is the process of *measurement* that allows us to get information from behind the curtain, and measurement in quantum mechanics is peculiar, just as peculiar as being in more than one state at a time. If you try to measure the polarization of a photon that is in a superposition of the states $|0\rangle$ and $|1\rangle$, in equal proportions, say, the result will be a classical number, 0 or 1, but it will be produced by throwing the dice that bothered Einstein so much. In fact, the result of a measurement will be $|0\rangle$ half the time and $|1\rangle$ half the time. Furthermore, *after* the measurement the photon will be in the "pure" state corresponding to the result of that particular measurement, either $|0\rangle$ or $|1\rangle$. We then say the photon's state has *collapsed*.

We are now in a position to describe the trick behind the quantum computer that solves Deutsch's problem in one shot. The mysterious information processing takes place, of course, behind the curtain. A quantum computer is built that incorporates the black box X and that operates on the superposition of $|0\rangle$ and $|1\rangle$, formed from the original, classical inputs 0 and 1. The key is that the quantum computer, in operating on the superposition state, processes both the $|0\rangle$ part and the $|1\rangle$ part of its input *simultaneously*, a piece of sorcery called *quantum parallelism*. The answer to Deutsch's question is extracted by a carefully designed measurement.[21]

The problem described in Deutsch and Jozsa (1992) actually deals with a more general problem, involving N bits, and the corresponding abstract space behind the curtain is of very high dimension—exponentially high. To see how this comes about, suppose we have two photons, each in the quantum states $|0\rangle$ or $|1\rangle$, or some combination of them. In this case the abstract state behind the curtain is a superposition of *four* possibilities (called the *basis*), $|00\rangle$, $|01\rangle$, $|10\rangle$, and $|11\rangle$, corresponding to the four possibilities for the pure states of the two photons. If there are three photons, there will be eight such possibilities, $|000\rangle$, $|001\rangle$, $|011\rangle$, and so on, since there are two possibilities for each slot. That's a total of 2^N; so with one hundred photons the space has dimension 2^{100}, which is about 10^{30}, an intriguingly large amount of parallelism that appears to be available for computation.

This picture, drawn with admittedly broad brushstrokes, is essentially how quantum computing works. We arrive at the very interesting question of how far this can be pushed.

11.7 Quantum Hacking

The God who made the world quantum mechanical not only plays with dice but, from the point of view of the computer scientist, giveth and taketh away. She giveth hope with quantum parallelism. But it is with the measurement process that God taketh away; what we can accomplish with a quantum computer depends on just how much information we can inveigle from the abstract space behind the curtain.

We've seen that the space behind the quantum curtain is of very high dimension—*exponentially* high. This does mean that we can operate (behind the curtain) on an exponential number of states at once, and suggests that we might be able to exploit this quantum parallelism to solve NP-complete problems in polynomial time. For example, we might be able to explore all the possible tours of a traveling salesman problem—simultaneously—and pluck the solution from behind the curtain with a cunningly designed measurement.

A breakthrough in this direction came in 1994, when Peter Shor published an astonishing quantum algorithm for a problem that is at the heart of our best methods of encryption, thereby getting about as much attention as an algorithm can get.[22] At this point quantum computing graduated from an enticing theoretical possibility to an entire field suddenly important to our national, corporate, and personal security.

The widely used RSA algorithm for public-key encryption is based on the apparent difficulty of factoring the product of two large prime numbers.[23] The best-known algorithm for this problem on a classical (non-quantum-mechanical) computer is exponential, and it is widely believed that there is no polynomial classical algorithm for the problem, although this has not been proved. It is also believed to be outside the class of NP-complete problems, although, again, this has not been proved. What is amazing about Shor's (quantum) algorithm is that it is polynomial! Thus, a quantum computer can break RSA encryption, and building quantum computers is now, naturally, funded lavishly by government agencies. Beyond that, the work has blossomed into the new and fertile field called *quantum information science*, important to physics as well as computer science.

11.8 The Power of Quantum Computers

As mentioned, factoring the product of two large primes *appears* to be difficult with classical computers, but is definitely possible in polynomial time with quantum computers. This immediately suggests that quantum computers might be able to solve NP-complete problems in polynomial time, which would mean that quantum computers would transport us to the promised land of computing, since it is widely believed that P ≠ NP and hence that the NP-complete problems are truly intractable for classical machines.

Before bursting this particular bubble, the phrases "appears to be difficult," "widely believed," and so on need to clarified. They come about in computer science, and science in general, when a claim has not been proved rigorously, but when the evidence has piled up. The nature of the evidence depends on the particular field. For example, many smart computer scientists and their often hungry graduate students have long been trying to achieve fame and fortune by finding an efficient (classical) algorithm for any one of the thousands of NP-complete problems. There is also theoretical evidence, based on using certain kinds of black boxes (oracles), that the P = NP question is in some sense deep. The oracle results also point to the result we expect, that P $\neq$ NP.[24] The pile of evidence is now high enough to convince most computer scientists that P $\neq$ NP, but it is also true that some of the most highly respected researchers continue to emphasize the fact that the door remains open, if only a crack. We should always remember that science is full of surprises.

Returning to the promise of quantum computers, the evidence is also mounting that the quantum computer, while provably outclassing classical computers on some problems, is not going to crack all the problems in NP.[25] Quantum parallelism itself is alive and well, but, apparently, the answers just can't be extracted using measurements with enough finesse to solve the problems we have reason to believe are the *most* difficult.

11.9 Life Itself

Living things process information in many ways, using both analog and digital representations. For one example, our metabolic rate is controlled by hormones produced in the pituitary gland, a control system using analog signals. For another example, the central dogma of molecular biology describes the transmission of information from DNA to messenger RNA to protein—all strictly digital.[26] The brain, the living thing closest to a recognizable computer, processes information in both digital and analog form.

And so we come, finally, to ourselves, and before going on we should run through a routine homework exercise: show that the brain is at least as powerful as a Turing machine. Solution: figure 11.3 shows a sketch of a neuron, the type of cell that is responsible for the information processing in the brain—the brain's "transistor." It receives input signals and produces an output signal; the details are many and vary greatly from neuron type to neuron type. What is important is that the signals appear at synapses of two types: excitatory and inhibitory. The former tend to promote the production of an output from the neuron, and the latter tend to block output. In the simplest case, a single inhibitory

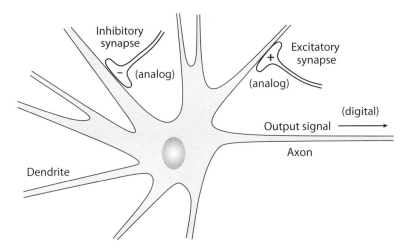

FIGURE 11.3. The neuron as valve. The synapse labeled "−" at the left is inhibitory and turns off the output; the synapse labeled "+" at the right is excitatory and will produce an output when the inhibitory input is off. Both synapses are part of the analog neuron, but the output signal, traveling to the right along the axon, is digital. The neuron can thus function as a valve, proving that the brain is at least as powerful as a Turing machine.

input can prevent a neuron from producing an output (firing), and we arrive at the valve of chapter 3, a universal building block from which we can build a Turing machine.[27] In fact, because no exponential blowup is involved in piecing together a general computer from valves, the brain can simulate a Turing machine with polynomial resources (time and hardware). This should come as no surprise, since Turing's conception was motivated by what a human can do with pencil and paper.

The reverse of this little exercise above would ask whether a Turing machine can simulate the brain, which would be implied, of course, by the Church-Turing thesis. If we ask for an *efficient* simulation of the brain, that would be implied by the extended Church-Turing thesis.[28] There are some who believe that there is something special about the brain, that its workings cannot be simulated by any kind of Turing machine, and that, therefore, some version of the Church-Turing thesis is false. My own view at this point is that this is tantamount to believing in magic, which is why I use the word in the title of this chapter.

The brain, the original personal computer, has the intriguing ability to direct the construction of computers—in effect, other brains. This self-reference leads inevitably to a heady idea, which has recently become fashionable: using our brains to design and build brains closes a feedback loop, which results in an exponential technological explosion. This general phenomenon— the runaway creation of more powerful brains by yet more powerful brains—is often called "the Singularity." The prospect is entertaining, of course, but highly speculative.[29]

11.10 The Uncertain Limits of Computation

In this chapter we tried to get beyond the power of the Turing machine. We seem to have come up almost empty-handed, with the exception of some truly extra zip provided by the quantum computer for some special kinds of problems like factoring. The best bet now seems to be that NP-complete problems are truly intractable for any kind of machine—in some deep sense tied to physical law.

In tracing the century-long arc from analog to digital, we progressed from principles that are quite firmly established (like the limitations imposed by noise and quantum mechanics), to conjectures in computational theory that are widely believed by specialists (like $P \neq NP$), to similar conjectures supported by lesser evidence (like the apparent limitations of quantum computing). The odd aspect of the appraisals of difficulty in this chapter is that they are tentative and based on conjectures. It is possible, although apparently unlikely, that $P = NP$ and that there is nothing difficult about NP-complete problems after all! Likewise, it may be that quantum computers can solve NP-complete problems efficiently; or that it is impossible to simulate what the brain does with a Turing machine.

With the possibility of the Singularity and the possible special nature of the brain, we have come to the end of well-charted territory. In the next, final, chapter we discuss how the principles of the discrete revolution have led us to today's internet-dominated world, and exactly what kind of singularity we can expect to see in the development of intelligent machines. In the process we return to the questions raised here about the computational power of the human brain.

Part IV

Today and Tomorrow

12 The Internet, Then the Robots

12.1 Ideas

To this point we've followed a path that is more or less historical, but guided principally by a succession of fundamental ideas. These ideas have led to today's wave, the internet—and will lead, in my view, to tomorrow's wave: the accelerating development of artificial intelligence and, inevitably, to autonomous robots, the androids of science fiction. This is a lot of territory to summarize in a short farewell chapter, but we have the advantage of the scaffolding provided by those few, relatively simple ideas. They led very naturally to today's world, and will, I believe, lead as naturally to tomorrow's.

Return to our point of departure, 1939, the analog world on the eve of World War II and, incidentally, the time of your author's arrival on this planet. The succeeding decade saw the birth of truly practical digital computers, assembled from thousands of hot electronic valves, a fin de siècle invention, using the nineteenth-century algebra of Boole. Those room-sized, sluggish beasts evolved and spread from scientific and business establishments to your pocket, by the grace of quantum mechanics, semiconductors, and the "room at the bottom" to which we were invited by Richard Feynman and Gordon Moore.

As Moore's law progressed, the sampling principle of Nyquist filled the screens of computers with colorful images and drove their loudspeakers with the voices and music that are the sounds of civilization. And following the laws and limits prescribed by Claude Shannon's beautiful theory of information, computers began talking to each other, until today the cultural globe is digital, enmeshed with the information networks that we recognize as the internet. Nourishing all of this flowering are algorithms— embodied in the programs of our billions of stored-program machines.

By my count, six basic ideas underlay the transformation from analog to digital that we have described. To take stock, following the progression in this book:

- **Signal standardization and restoration** protects information from being destroyed by noise. The principle defines what it means for computation to be digital and is incorporated in the conceptual machines of Babbage and Turing. But, as discussed earlier, we should not be ready to write off analog computation quite yet. There may, after all, be power hidden in the analog aspect of the physical world that will ultimately prove important, perhaps even decisive.

- **Valves** allow one standardized signal to control another and, together with fan-out, are by themselves enough to implement any logical operations. Historically, they have been realized with electromagnetic relays, then electrons moving in a vacuum (vacuum tubes), then electrons moving in a semiconductor (transistors). The idea of implementing logic with valves is founded on the mid-nineteenth-century mathematics of George Boole.

- **Moore's law** is possible because the universe has very fine granularity. As Feynman observed, there is plenty of room at the bottom. The operation of Moore's law for at least five decades resulted directly in the proliferation of the personal computer.

- **Nyquist's sampling principle** ensures that if we sample audio and video fast enough we can mirror any analog signal processing with digital signal processing.

- **Shannon's noisy coding theorem** shows that it is possible to achieve (essentially) noise-free digital communication, provided that we accept the limitation of bandwidth and the delay and computational cost of encoding and decoding. Myriad personal computers are now connected via the internet. Shannon's theory defines the nature and limitations of bandwidth.

- The **Turing machine** exemplifies the stored-program, conditional-execution digital machine. Any computer you use today is in principle no more powerful than a Turing machine, and what it does is determined by the programs it runs.

Today the last two of these ideas—really just digital *communication* and *computation*—are driving the winds of change. First, the abundant communication provided by the internet is changing and challenging human society and culture. People have become astonishingly interdependent on a global scale; Earth has become thoroughly soaked in information-bearing signals. Second, algorithms—for hiding and stealing information, for solving problems of biology and physics, and for mimicking thought itself—are becoming our most powerful tools (and weapons!). Next, we look at the internet from our particular point of view: What are the most fundamental ideas that make the internet possible?

12.2 The Internet: Packets, Not Circuits

Here is a direct effect that the discrete nature of information has on your life. Your computer can be connected to any one of, say, a billion other computers—in the blink of an eye. How would you design a system of connections that can make such an amazing thing happen? If you follow the analogy with telephone connections, you would seek out a path from your computer to your target destination, fix the path for the duration of the connection, and exchange information along this path. This method is called *circuit switching*. For example, to route an old-fashioned telephone call from New York to Hong Kong, we might find a connection that goes from New York, to Chicago, to Los Angeles, to Sydney, to Hong Kong. Once that path is established, it would be used for your entire call.

But the fact that information today is almost always available in digital form, especially when you are using a browser, means that we can do things in a completely different way: we can break your signal in pieces, called *packets*. The packet contains a piece of your data, but it also contains a lot of information in its *header* and *trailer* : the packet's length, its source, its destination, its "time to live" (before exceeding the maximum number of hops it is allowed to take before it dies), a checksum (for detecting errors), an identification tag that is used to assemble the packet together with other packets to reconstruct the original message, and so on. Now each of these packets leaves your location and goes hopping around, from point to point, looking for a path to your destination. It is quite possible that the packets in your original message reach their destinations via many different paths. It might be that some packets in a connection from you to Hong Kong use Seattle as an intermediate node, others might use El Paso or, for all we know, an orbiting satellite.

The most obvious advantage of packet switching over circuit switching follows directly from the fact that we can break the message up into small packets. Any particular link in the path of any particular packet may be shared with many other packets that are part of many other messages. So if many people are sending many messages to many other people, all at the same time, the communication links in the network are used much more efficiently than they would be if dedicated circuits were used. Think of how often you are simply not typing or downloading—why tie up a dedicated circuit with idle time?

There are other advantages to packet switching, but there are also situations where circuit switching is better. Packet switching is generally more resilient to network failures: if a packet is lost or dropped because it happens to get stuck somewhere, it is easy for the receiving node to learn this and request a retransmission of the missing packet. On the other hand, packet switching may incur more delay than circuit switching because once a circuit is established, the transmission can proceed at full speed. This may be a critical constraint in situations where delay is not tolerable; as an example, consider a surgeon performing a delicate operation at a location remote from the patient.

By and large, though, packet switching is an enormous win for the internet and the digital idea, since the digital form of data makes it easy and natural to put into practice. Simply put, breaking our messages into small pieces—which is much easier for digital, as opposed to analog, signals—makes it possible to make much more efficient and reliable use of our channels.

12.3 The Internet: Photons, Not Electrons

We think of "wireless" today as meaning that radio is used instead of copper wires. Don't forget, however, that people have been communicating long-distance without wires using light signals for thousands of years, using smoke signals by day and fires by night.[1] The early 1800s saw the development of the Chappe semaphore, which sent signals between towers on the tops of hills, 5 or 10 km apart. One such chain of 220 towers extended from the Prussian border via Warsaw to St. Petersburg. This was the state of the art in long-distance communication when Alexander Graham Bell and his assistant Sumner Tainter became obsessed with the idea of sending voice with light beams. They were faced with the formidable problems of imprinting signals on a light beam (modulation) and detecting the variations (demodulation), but the birth of the *photophone* was documented on February 19, 1880, with the message "The problem of the reproduction of speech by the agency of light was solved by Mr. Sumner Tainter and myself in my laboratory..."—about 20 years before the successful transmission of voice by radio.[2]

The idea of communicating using light instead of electricity was reborn with a vengeance in the 1970s, and fiber optics has fueled the explosion of the internet. Today our streets are being dug up and bundles of hair-thin optical fibers laid down to connect us to everywhere, at speeds that were unthinkable just a few years ago. The number of bits that can be pushed through an optical fiber in a second has, in fact, increased exponentially, following its own kind of Moore's law. Figure 12.1 shows the progress in optical fiber speed over the last three decades or so. Hecht (2016) proposes the name *Keck's law* for this optical version of Moore's law.

We now need to ask a really basic question: Why does long-distance transmission of information using photons in glass beat electrons in copper? The answer lies in the problem of *loss* and the *skin effect*. When a signal of any kind propagates down a wire or fiber, there is an inevitable loss in its size. Given that there is always a certain amount of noise, this limits the distance the signal can travel. However, digital signals can be regenerated by

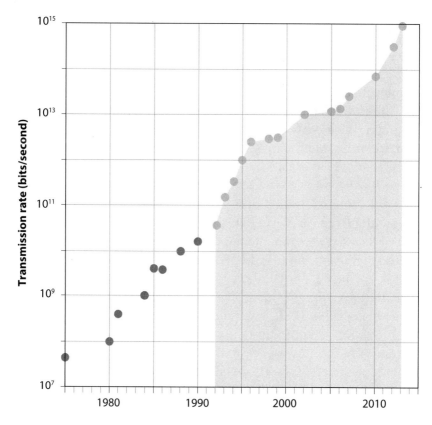

FIGURE 12.1. The progress in the speed of optical fibers. The vertical axis is the rate in bits per second in the most aggressive experiments. The shaded area indicates the use of *wavelength-division multiplexing*, in which several different signals are sent down the same fiber simultaneously, using different wavelengths for each. See Hecht (2016) for a review of the progress in fiber optics technology over the last few decades. He proposes the name *Keck's law* in analogy to Moore's law, after Donald Keck, the coinventor of low-loss optical fiber. (Available at https://www.eitdigital.eu/news-events/blog/article /after-moores-also-kecks-law-looks-in-trouble/. Accessed September 15, 2017. Courtesy IEEE. Great leaps of light. *IEEE Spectrum*, 53 (2): 28–53, February 2016. Reprinted with permission.)

deciding on the zeros and ones before the noise masks them and then generating a brand-new signal with a refreshingly large size. The devices that do this regeneration, called *repeaters*, are not cheap, however, and installing them in undersea cables is especially troublesome. Therefore, the smaller the loss, the more practical the long haul of signals.

This is where optical fibers shine in comparison with copper. When high frequencies or short pulses travel through a conductor like copper, the electrons tend to concentrate near the surface—the "skin"—of the conductor. This makes a copper wire seem to have a much higher resistance, since the effective diameter of the wire is much smaller. Signals are therefore attenuated more quickly, and the faster the pulse speed, the more the loss. The skin effect simply does not occur when photons travel through glass fiber, and the development of very low-loss optical fiber has been a boon to the growth of the internet. Using the terminology of chapter 7, optical fibers have, for a given cost, much more *bandwidth*—in the sense of Shannon information theory—than copper wires.

Glass fiber also has several other advantages over copper: It is immune to electromagnetic interference, which includes radio signals and noise from electrical equipment. It offers some built-in amplification using small amounts of rare-earth erbium, and this amplification can greatly extend the range of fiber transmission with no need for repeaters. It is also more durable, lighter, and, in the long run, cheaper.

The real advantages of glass fiber stem from the differences in the physical properties of photons and electrons. Hecht (2016) puts it very well: electrons interact strongly with other matter and are therefore well suited for logic and memory; photons do not interact strongly and are perfect for long-distance communication—where interaction is highly undesirable. When the time was ripe, we saw the exponential explosion of electron-based chip technology and several decades later the explosion of photon-based fiber transmission. Today we enjoy the benefits of both growth periods.

Consequences

The incredible blossoming of the internet can thus be traced largely to the two fundamental factors just discussed: packet switching and the development of optical fibers. As we know, however, the net presents us with great dangers as well as great opportunities. When computers sat as isolated machines in the corner of a laboratory or study, life seemed simple. Data was hard to come by. Not many texts or books were available in digital form, data was precious, and shared, if at all, in small communities. It may be hard to imagine today, but a program that ran on one machine did not necessarily run on another. People and their computers generally kept their digital noses out of other people's business.

With the internet, this serene picture changed very quickly and drastically. Incredibly cheap and fast digital communication and, of course, the chips of Moore's law have led to oceans of data, and ubiquitous computing. For one thing, it isn't always necessary to own your own computer. If you are starting a business and don't want to invest in hardware just yet, you can ship your computation—and storage—to a roomful of machines at some, as we say, undisclosed location, picturesquely referred to as "the cloud." When it is so cheap and fast to send billions of bytes back and forth across the continent, why worry about buying and maintaining all the machines you need to run a growing business?

With all the data flowing around us, all fitting snugly in its allotted bandwidth according to Shannon's theorem, the inevitable happens: it gets collected, and it gets exploited—by businesses for marketing, by medicine for good, and by criminals for evil. The phenomenon is collected under the rubric of "Big Data," and we all know today that someone, somewhere may be watching our keystrokes, hopefully without our names and social security numbers attached.

Cloud computing and big data are both consequences of the abundance of communication and memory, and these follow from the ideas we associate with the names Moore, Nyquist, and Shannon. The great dangers of the internet are intimately associated with the name Turing, who pioneered both the conceptual basis for programming and the use of the stored program for locking up and burglarizing information. As we've seen in the preceding chapter, the difficulty of cracking encrypted information is tied to the most fundamental and difficult theoretical questions in computer science, and supplies the most urgent pressure for developing quantum-based computers.

In fact, the idea of the stored program is so general and so powerful that it has insidious consequences. Your computer, as you know full well if you live in the twenty-first century, is vulnerable to invasion and subversion by viruses—just as is the reproductive mechanism of the living cell, and for much the same reasons. To put it a bit grimly, we are witnessing today an arms race between the innocent consumers of our lovely digital technology and the evil hackers. Just check your spam bucket carefully.

It is all playing out from our six ideas, plus some packet switching, some optical fiber, and a few billion lines of code.

12.4 Enter Artificial Intelligence

I believe it is not hard to see where the six ideas are leading us: Just check the Technews website of the Association of Computing Machinery (ACM), which regularly posts news of the computer world.[3] The snippets and reports of research from around the world fall into well-defined categories, and dominant among them are new applications and algorithms for what is generally termed *artificial intelligence* (AI),[4] and new kinds of materials for fabricating gates (in the last analysis, valves) and sensors. These are the makings of artificial minds and bodies: we are headed toward independent, stand-alone, often called *autonomous*, robots—or, as we termed them at the beginning of this chapter, *androids*.

Some comments about terminology, which in this field changes fast and is sometimes not clear: "AI" is rather old and dated terminology, today perhaps too vague to be useful to researchers, but very popular in general discourse. Maybe more fashionable in the computer science crowd is the narrower term *machine learning*, the ability of a computer to successively refine its algorithm based on feedback from application to a task. Machine learning systems can be considered a subset of AI. Further specialized are systems lumped under the term *connectionist*, which means that the machine's algorithm for intelligent behavior uses a network of interconnected, simple, often identical units. When those units are intended to behave roughly like neurons, we have an even more specialized class of systems, termed *neuromorphic*, which are also called *neural nets*. I restrict discussion here to this special class of AI systems, actually, today, perhaps the most promising and successful.

Neural nets mimic, albeit very crudely, the way neurons might be organized in the brain:[5] there is a set of input (artificial) neurons, these connect to another layer of neurons, and so on, until we reach the final output layer, from which results are taken. Getting the results is, of course, the object of implementing the neural net in the first place, usually by simulating it with a general-purpose computer. For example, a system designed to recognize speech, one of the popular applications of neural nets, will have inputs from a microphone, perhaps processed by filters covering different bands of frequencies; and its output will be words in text form. We already have software that can do this job, but it isn't as good as humans. Naturally, the aim in designing neural nets is, often, to match or beat human performance.

12.5 Deep Learning

The first, simple neural nets were originally constructed with one input layer of (artificial) neurons,[6] one intermediate layer, and an output layer. What is not known at the start are the *weights* assigned to the connections between neurons. Usually, these weights can be any number between −1 and +1, with 0 representing no connection at all and +1 or −1 representing the strongest possible connection. The weights therefore determine which neurons are connected to which, and finding ("learning") good weights for a particular task is the major computational challenge in using neural nets. This process is usually called *training* the neural net.

Training neural nets, as well as picking the right kinds of neurons in the first place, is both an art and a science and is the focus of intense research as I write this. This should not be surprising. Humans, for example, are born with about 100 billion neurons and a certain number of synapses (roughly speaking, connections), many in a completely untrained state. A baby— or rather a baby's brain—learns how to connect those neurons (adding synapses) to recognize its mother's voice, focus both its eyes on one object, pick things up, and walk. This is not to mention advanced skills like understanding spoken language, speaking in coherent sentences, and sending text messages. As we know, further training those 100 billion neurons to the point where the individual can drive a car sensibly,[7] empathize with others,[8] and make and train babies with their own 100 billion neurons takes a couple of decades. Some people never quite get there. Analogously, progress in the development of neural nets—by introducing more neurons, or more than one intermediate level of neurons, for example—has been limited by the time required for training. For some complex tasks, the learning process is now simply beyond the capabilities of available computers.

The idea of neural nets has tended to pop up repeatedly, in one form or another, over the last century.[9] Their repeated appearances were attended by overly enthusiastic promotion and exaggerated claims, to be followed by disappointment and fading. The inexorable progression of Moore's law changed all that. By the 1990s our little chips had become much, much faster, and architectures were developed for putting many of them to work at the same time. It became possible to build true "supercomputers." By the beginning of the twenty-first century, it became practical to build and teach neural nets with many intermediate layers, "deep neural nets," and the field of "deep learning" was born. As I write this, they are all the rage, attracting many gifted researchers and corresponding resources. And today, deep neural nets are often the best tool available for tasks like computer vision, handwriting recognition, and natural language understanding. These are applications that are chosen in part because there are no clear-cut conventional algorithms that solve them. It is also true that humans are "designed" by evolution to be very good at these tasks, and it is perhaps not an accident that neural nets, which mimic, in a rudimentary way, the operation of the brain, are also good at these same tasks.

Skinner's pigeons

In 1940 B. F. Skinner, the famous behaviorist, had an idea for building what we now call "smart bombs."[10] Skinner was a great proponent of the idea of conditioning, and he started experiments to show that pigeons could be trained to peck at a moving image on a screen. The idea was that if an image of a target were projected on a screen inside a bomb, the movement of the head of a pecking pigeon harnessed inside the bomb could be made to control the direction of the bomb's flight.

The story of the ups and downs of Skinner's project makes entertaining reading, but I bring it up here because it illustrates very well the basic strategy of much of today's research in AI:[11] to make and use a useful neural net, replace the poor captive and doomed pigeon with a computer program or electronic circuit that reflects, very roughly, the way neurons interact in the pigeon's brain, and train it the same way—by reinforcing desired behavior with rewards.[12] In this case the reinforcing is accomplished by appropriately adjusting the weights on the connections between neurons, the synthetic synapses.

Deep learning is being applied today to many problems for which precise algorithms do not exist, like recognizing faces, reading sloppy handwriting, and understanding speech. Humans are very good at these jobs, as mentioned above, but only after years of training. And so it is not surprising that the real challenge in teaching neural nets to do useful things is the computation time needed to train them.

12.6 Obstacles

We've already noted that the history of neuromorphic computing, and AI in general, is marked by periodic surges of enthusiastic claims followed by disappointed expectations. Consider, for example, the June 2017 issue of the *IEEE Spectrum*, whose cover shows an array of rather cheap-looking human brains and the question, "Can We Copy the Brain?" The first paper in the special report section is titled, "The Dawn of the Real Thinking Machine."[13] The text illustrates some common speculative themes:

> Eventually, our tools will think for themselves, perhaps even becoming conscious. ... If our tools think for themselves, they could turn against us. What if, instead, we create machines that love us?

The last a pleasant thought. Perhaps.

Also in the same special report is an article that takes a much more cautious approach.[14] The author Gomes ends with the prediction, "[neuromorphic computing] will either take flight and soar over the chasm, or drop into obscurity." I would add a third possibility: The field will either take flight ... or go underground once more, to emerge 13 or 17 years later, like the periodical cicadas. I argue that if not now, eventually, the technology upon emergence will be advanced enough to support truly intelligent machines. By the way, Gomes uses the flight metaphor to remind his readers of the observation that successful aircraft do not flap their wings. An excellent point, I think: Perhaps the intelligent machines of the future will think using brains that are nothing like our own.

"Prediction is very difficult, especially about the future"—a quote often attributed to one or the other of two intellectual forces in the twentieth century: Niels Bohr and Yogi Berra. But the future of AI is so important that I will spend our remaining pages discussing some bright prospects for progress, some daunting obstacles, and possible consequences to society that are truly transformative.

Counting connections

Supporting Gomes's restrained attitude are some staggering numbers. I have already mentioned the 100 billion (10^{11}) neurons that constitute our endowment at birth—a generally accepted estimate. We must make do with those neurons, give or take a few, for our entire lives. We can't really complain of being shortchanged: after all, that also happens to be about the number of stars in the Milky Way. However, as large as the hundred billion may seem, it is tiny compared with the number of *connections*, the *synapses*.

Figure 12.2 shows a diagram of a typical neuron: drawn a little more realistically than figure 11.3, where we were concerned only with its ability to act as a valve. The synapses lie at the ends of branches from an axon that leaves the central part of the neuron, the *cell body* or *soma*. Each synapse communicates a signal from its neuron to another neuron, connecting across a small gap, to the other neuron's soma, or to one of its *dendrites*, branched

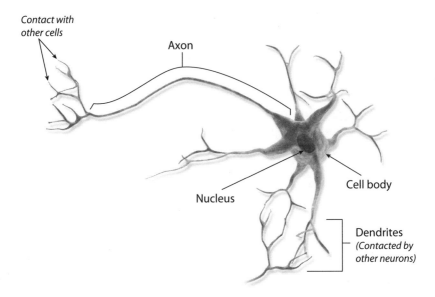

FIGURE 12.2. A diagrammatical representation of a typical neuron. The neuron's cell body collects signals through synapses from other neurons and produces an all-or-nothing response in the form of a train of spikes that travels along the neuron's axon (in this figure, to the left). The axon thus carries a *digital* signal, while the synapses are *analog*. Keep in mind that there are many diverse kinds of neurons, and each can communicate with other neurons via many, sometimes thousands of synapses. (Adapted from Blausen.com staff (2014).)

structures that collect signals from other neurons and transmit them to the soma. Generally speaking, with exceptions we can ignore here, the flow of information is from a neuron's soma, along an out-branching axon, to the synapses of other neurons. New synapses can appear, especially in babies, during learning and storing new information (the process we call *remembering*). Furthermore, the weights that are used by a neuron to combine its incoming information to form an outgoing signal can also be changed, much as the weights in a neural net can be changed during training. This means that the brain is constantly changing as we learn; it is continually rewiring itself. There may be as many as 10,000 synapses that belong to a given neuron, each one connecting to other neurons. Pakkenberg et al. (2003), who actually did a lot of counting, with a microscope, concluded that in the neocortex, where we do all our serious thinking, each neuron has about 7000 synapses for the exchange of information. Just to avoid messy numbers, let's say that the average for all neurons is 10,000, or 10^4. Each of 10^{11} neurons is therefore connected (by a one-way street) to, say, 10^4 other neurons. That makes on the order of $10^4 \times 10^{11} = 10^{15}$ connections in the human brain, a nice round number, a quadrillion.

These numbers are daunting when you consider trying to simulate the brain with a computer. Suppose the quadrillion connections correspond to weights that need to be learned in a neural net. Suppose we adjust the strength of each connection at each iteration of the training process, and we update the strengths at a 1 MHz rate, a million times a second.[15] That makes a billion seconds if we want to simulate the brain, or about 32 years—just to do one round of weight adjustments. And training for a reasonably sophisticated job can easily take thousands of rounds.

It is worse than that, because the neurons in the brain, together with their dendrites and synapses, are complex analog systems, governed by differential equations, and certainly not modeled accurately by the simplistic artificial neurons used in neural nets. Even if we use many processors in parallel, it is easy to see that a full and accurate simulation of the human brain is not feasible in the near future.

The bee brain

The industrious honeybee, however, brings us some good news. Its brain, the size of a single sesame seed, has only about a million neurons; human brains beat it 100,000 to 1. Furthermore, researchers have shown recently that the bee is capable of what is, for an insect, higher learning.[16] Bees can learn *concepts*, abstractions that are independent of a particular physical instance. For example, they can learn the ideas of "same," "different," "above/below," and "left/right," and apply them to different situations. Beyond that they can master two such concepts simultaneously.

The reduction in complexity in going from the human to the bee brain is even more dramatic than might be suggested by the reduction in the number of neurons by a factor of 10^5. Consider the number of *connections*, which tends to vary as the *square* of the number of neurons. The number of connections, a better measure of the complexity of a brain, therefore decreases by a factor of 10 billion (10^{10}). This is certainly encouraging. The example of the honeybee, an insect genius with a tiny brain, gives us hope that it might be possible to train neural nets of manageable size to perform relatively high-level tasks.

Is there analog magic in the brain?

The brain uses both digital and analog signal processing. Generally, interneuron communication is digital, using a coding of spikes that are sent along the threadlike axons that run between neurons. But the local operation at each neuron itself combines signals from other neurons in a very complicated way, using decidedly analog processing. As mentioned above, simulating a neuron accurately forces us to solve differential equations, which is a much more difficult and time-consuming computational job than implementing simple logic.

It seems that the brain uses digital encoding for interneuron communication for much the same reason that we use digital encoding for the internet. The digital form of the data sent from one neuron to another provides noise immunity, one of the six main ideas that explain why our entire civilization's information technology has become digital. Consider this: a typical neuron's soma is roughly 10 microns, or 10^{-5} meters in diameter, whereas the axon of our sciatic nerve is roughly 1 meter long, extending from the lower spine to our big toe. The signal along the axon of the sciatic nerve therefore travels 100,000 times farther than signals within the body of the neuron itself, which uses analog computation. Nature learned to use digital processing for long-distance communication, just as we have.

The analog processing in the brain's neurons returns us to a question about computational complexity raised in chapter 11. Does the brain use analog processing only as a matter of relative efficiency? Or does analog processing offer an exponential and hence qualitative advantage, thereby transcending the power of digital processing? As I mentioned in chapter 11, no one knows for sure, but the smart money is on the extended Church-Turing thesis: there is no analog magic going on in the brain (or anywhere else). If we accept this conclusion, then it is no real limitation in principle to try to emulate the brain with completely digital computation.

Is there quantum magic in the brain?

This still leaves quantum mechanics as another resource that the brain might use. We've seen in chapter 11 that quantum computers promise some important improvements in computational efficiency. Does the brain make use of quantum mechanics? This is a most interesting and provocative question.

As best we know, the brain, like any other material object, is *governed* by the laws of quantum mechanics. That is not at issue. The question is whether the brain takes advantage of quantum-mechanical effects in its computation.

I will only summarize the leading arguments pro and con. The most well-known proponent of the idea that quantum mechanics is used in an essential way in the brain is Roger Penrose, and the general proposal is put forward in *The Emperor's New Mind*, a brilliant if controversial work.[17] He develops some of the topics we covered here in more detail, and despite a sprinkling of equations here and there, the book is still suitable for the nontechnical reader. More specific proposals for the location and nature of quantum-mechanical computation in the brain are developed further in Hameroff (1994) and Hameroff and Penrose (1996). In particular, they propose that the source of consciousness can be traced to the quantum mechanics of microtubules, cylindrical protein lattices in the brain's neurons.

On the other side of the ledger, these proposals have generally been met with skepticism by biologists and physicists. Quantum systems, which store information in quantum states, are very delicate. They tend to interact with their environment, thereby losing quantum information, a phenomenon called *decoherence*. This loss is the main technical difficulty in doing any kind of practical quantum computation, in the brain or not. Critics argue that keeping quantum states protected from decoherence in the brain is implausible: the brain is just too wet and too warm.[18]

We have just mentioned three possible obstacles to building an artificial brain for a robot: the sheer complexity of the genuine article, and the possibilities that it might require analog computation or quantum computation—thereby acquiring power that goes beyond the standard Turing machine. None of these obstacles is necessarily fatal. If history is any guide, we will be packing more and more computation of some kind into smaller and smaller virtual skulls, and if necessary we can always equip the robot brain with any analog or quantum-mechanical features that might be required. After all, we have a proof of principle: ourselves.

12.7 Enter Robots

Who is learning what?

Skinner trained his pigeons to follow a target on a screen. I am sure he was an expert at training birds, but what did he learn about the algorithms that they use for pecking at a dot on a screen? In the early 1940s there were, for all practical purposes, no digital computers on earth, and only Alan Turing —plus perhaps a handful of others—were even thinking about algorithms and corresponding programs for computers. To learn what the pigeons were learning would have required understanding how the computer we call the pigeon brain works, and 75 years later we are still a long way from that.

The situation today in AI is closely analogous. After training a deep neural net to do something like transcribing handwriting, the computer scientist who built the net and trained it usually knows very little more about the problem of recognizing handwriting than when she started: the result of training is simply a large collection of numbers for the weights in the net, reflecting how the artificial neurons are interconnected, the artificial synapses. And as neural nets get deeper and more complex, the problem becomes more like that of understanding how biological brains work. Today understanding how deep neural nets do what they do is an important research area. So far, in practice, using neural nets to "solve" a problem amounts to delegating its solution to another thinker. Which is a way of introducing the robots.

Čapek and Dick

Given technology's track record, especially over the past century, it seems inevitable, to me at least, that we are headed quickly toward the creation of humanoid robots—a step that has been widely anticipated in science fiction and that by now has thoroughly invaded the public consciousness.

Karel Čapek introduced the term *robot* in his play *R.U.R. (Rossum's Universal Robots)*, written in 1920 and premiered the next year in Prague.[19] Robots are now also called *androids*, or in P. K. Dick's classic *Do Androids Dream of Electric Sheep?, andys.*[20]

Čapek, with his unique genius and characteristic flair, raised, in their very first appearance, the central question about robots: How human are they? Domain, the general manager of Rossum's Universal Robots, declares, in the first act of Čapek's play:

> DOMAIN. Mechanically they are more perfect than we are, they have an enormously developed intelligence, but they have no soul.

Later in the same act, Helman, the psychologist in chief, responds to Helena as follows:

> HELMAN. They've no will of their own. No passion. No soul.
> HELENA. No love, no desire to resist?
> HELMAN. Rather not. Robots don't love. Not even themselves.

Helena, a visitor to the R.U.R. factory, is Čapek's voice of humanism. A bit further on we learn more.

> HELMAN. Occasionally they seem somehow to go off their heads. Something like epilepsy, you know. We call it Robot's cramp. ... It's evidently some breakdown in the mechanism.
> DOMAIN. A flaw in the works. It'll have to be removed.
> HELENA. No, no, that's the soul.

Later on the question of robot pain is raised, and then of love: inescapable elements of the experience of being human.

Almost 50 years later, P. K. Dick approaches the same question in his own way. His protagonist, Rick Deckard, is a bounty hunter whose assignment is to track down and retire (kill) andys, who have escaped from Mars. The escaped andys are an advanced model, using the Nexus-6 brain unit, and it is possible to distinguish them from humans only with the Voigt-Kampff empathy test: for Dick it is empathy that is the essentially human feeling.

Passion, the soul, love, pain, empathy: They are all manifestations of the peculiar thing we call *consciousness*.

12.8 The Problem of Consciousness

The hard problem

Is a robot, a machine that mimics the thinking of humans in one way or another, *conscious* ? This is another way of asking, as Helena did in *R.U.R.*, whether robots have a soul, say, or feel pain, or love. For that matter, why do the brain processes of *humans* result in subjective experience?[21] How does the physical operation of neurons result in our seeing the color red? Or feeling the pain of a toothache? This question is what philosopher David Chalmers calls the *hard problem* of consciousness.[22] It is of a different order of difficulty from other questions we might consider. Chalmers puts it this way in the preface to his book *The Conscious Mind*:[23]

> Consciousness is the biggest mystery. It is probably the largest outstanding obstacle in our quest for a scientific understanding of the universe. ... how could a physical system such as a brain also be an experiencer?

One way to deal with the hard problem is to surrender. This is the response of the *mysterians*.[24] Their position is simply that we, with human brains, are fundamentally incapable of ever understanding what gives rise to the subjective experience of consciousness. This idea is foreshadowed by a remark of the great biochemist Jacques Monod:[25]

> The logician might be moved to remind the biologist that his efforts to "understand" the entire functioning of the human brain are ordained to failure, since no logical system can produce an integral description of its own structure.

Whether this sharpens the argument or further confuses matters, I leave to the reader.

Strong AI

John Searle considers quite a different response to the problem of consciousness, which he calls *strong AI*:[26] the view that any computer that runs the right program will be "conscious" in the same way the brain is. As simple as that. This is to claim that the brain has no secret ingredients like quantum mechanics in neuron microtubules or "pixie dust,"[27] or, for that matter, any of what Chalmers lumps into the category "new physics."[28]

Searle introduced the term *strong AI* in putting forward what is generally considered the best argument against it, which has become known as the "Chinese room" argument.[29] It is the following thought experiment:[30] Suppose we start with a computer program that, in the spirit of strong AI, incorporates some aspect of consciousness, such as understanding Chinese. Lock a non-Chinese-speaking demon in a room, who will read through the program and execute it using slips of paper, one instruction at a time. Never mind that this will entail following millions or billions of instructions, and never mind that the demon does not know what the symbols for Chinese characters mean; this is a thought experiment. The demon is prepared to receive questions written in Chinese characters, which it doesn't understand at all, look up the corresponding instructions of the program, and return answers, also written in Chinese characters. Chinese speakers arrive at the room, drop questions written in Chinese through a slot in the door, and the demon returns answers, also written in Chinese characters.

Now, as far as the Chinese visitors are concerned, the Chinese room understands Chinese. But the demon knows absolutely nothing of the Chinese language. Thus, the argument goes, executing a program is not sufficient for understanding Chinese, or, in fact, understanding anything. Searle argues in this way that, in general, executing a program cannot give rise to consciousness.

Defenders of strong AI respond that it is not the *demon* that knows Chinese but the *system* of room plus program plus demon. Chalmers sees this point as an impasse, with proponents of strong AI concluding that the system is conscious and opponents finding the conclusion ridiculous. Chalmers actually goes on to argue for breaking the impasse in favor of strong AI. But we must leave the problem of consciousness now, in this most unsatisfactory unanswered state. Be assured, it is too interesting and important a problem to go away.

12.9 The Question of Values

Despite the talk of dangerous androids who have escaped from Martian colonies, and hypothetical Chinese rooms, these arguments have important consequences. The answers to our questions matter. Suppose, for example, that machines can be conscious and, further, can feel pain and suffer. Then we, as humans, have a moral responsibility, at the very least, to consider that suffering.[31] We, our culture and our genes, are the product of a few billion years of Darwinian selection, and this accounts for that moral responsibility, as well as our own soul, love, pain, and empathy.

On the other hand, we will design robots from scratch. It is up to us to provide them with a moral compass. Their track record is not encouraging, but that may be because we expect a great deal of action and conflict from our science fiction. Čapek's robots revolted and replaced humans. Dick's andys were dangerous indeed, and with the distinction of lacking empathy entirely. Isaac Asimov addressed the problem in his story "Runaround" with his "Three Laws of Robotics":[32]

1. A robot may not injure a human being, or, through inaction, allow a human being to come to harm.
2. A robot must obey the orders given it by human beings except where such orders would conflict with the First Law.

3. A robot must protect its own existence as long as such protection does not conflict with the First or Second Laws.

A good beginning, but not sufficient, I would say, to help posthumans enjoy anything like our cultural heritage. I would implore the robot makers of the future to consider them our children—to lavish on them the same attention to moral and artistic education that we lavish on our biological offspring. They may bear the responsibility for carrying our cultural heritage and values into the mysterious centuries ahead. We want to make them proud of their human origins and ensure that they preserve our peculiar human values.

To consider our most fulfilling enjoyments: there is love, of course, and art. But it is a sad fact, I think, that there seem to be so few fundamental scientific laws to discover. Once they are discovered, the fun is gone. Newton got to discover the elegant way in which the motion of all bodies is governed by a simple gravitational law, and in doing so, he spoiled it for the rest of his slower colleagues. And Einstein, in finding both special and general relativity, not to mention explaining the photoelectric effect, was, if I may be forgiven, a bit greedy.

Stravinsky, on the other hand, was not really depriving anyone of the opportunity to compose, say, his *Rite of Spring*. The likelihood of anyone independently coming up with that particular composition, or anything like it, is infinitesimal. The same can be said, of course, for all the great painters, composers, writers, and artists of any kind. Their creative work simply cannot be viewed as entries in a competitive race to any particular goal—there are simply too many possible paintings, operas, symphonies, or novels to worry about the problem of competition. In this sense, art is, in the long run, a lot better than science. We and our progeny—analog, digital, or hybrid; human, android, or a society of both—will never run out of art to create and enjoy.

Having reached, and perhaps passed, the boundaries of fair speculation, it is time to bid the reader a cheerful farewell. I do, however, wish to leave you with the contents of the following intercepted subspace transmission, just in.

Epilogue

Interception

It's a message coming from the Gamma 73 sector, from some boring carbon-based bipeds. Looks like they're offering some stuff for sale ... here, they're taking bids on some kind of digital, electronic robot.

[*Laughter from the group gathered in the receiving salon*]

Hmm, here's some stuff of possible interest ... the genetic code for an animal they call a "cat," possibly similar to our custom personal guides but without empathic or language skills.

[*More laughter*]

What have we here? Something about a magical instrument called a "flute." Written by someone they call "Mozart." It seems to be in the form they call "audio," which is a frequency range we will need to translate up.

[*With renewed attention the group downloads and listens to the short sample, heterodyned to their hearing range—after that, stunned silence.*]

Wonderful! We'll want more of *that*. With the authorization of the Council, we will offer to trade plans for, say, a gravitonic robot for the full *Magic Flute*.

[*Cheers*]

It looks as if we have some valuable trading partners in the Gamma 73 sector after all.

Notes

Chapter 1 / The Discrete Revolution

1. Leaving aside some pioneering sparks in the dark, like Charles Babbage's engines, in the 1830s and 1840s. We return to Babbage in chapter 9.
2. It took a bit of negotiation with my friends to help me lug these behemoths home, but I had the luxury of a secret basement laboratory to house my treasures.
3. Khan (1991).
4. In Basdevant (2007), p. 6, a personal, and delightful, introductory treatment of quantum mechanics. Be warned, it is mathematical, being a record of his lectures at the École Polytechnique.

Chapter 2 / What's Wrong with Analog?

1. Note that distortion is also a form of noise corruption.
2. Also called *Johnson noise* because it was discovered and first measured by B. J. Johnson; see Johnson (1928). H. Nyquist explained the phenomenon using thermodynamics and statistical mechanics, in a paper that followed Johnson's in the same issue of the journal; see Nyquist (1928b). It takes both experimentalists and theoreticians to make science work.
3. For some interesting history, a description of Einstein's derivation, and a description of a reproduction of Perrin's experiment using modern equipment (a charge-coupled detector, or CCD camera), see Newburgh et al. (2006).
4. At the risk of assuming too little, the metric prefixes pico-, nano-, micro-, mega-, giga-, and tera- mean, respectively, trillionth, billionth, millionth, million, billion, and trillion, which in scientific notation are factors of 10^{-12}, 10^{-9}, 10^{-6}, 10^6, 10^9, and 10^{12}.
5. As put by Horowitz and Hill (1980), the bible of practical electronic design.
6. The relative size of shot noise also depends on the range of frequencies present in the particular signal involved; in Horowitz and Hill's example, 10,000 Hz, a bandwidth suitable for high-quality telephones. A discussion of the frequency content of signals follows.
7. Press (1978) was interested in how quasars can radiate such stupendous amounts of energy. As it turns out, more progress has been made in understanding the power of quasars than $1/f$ noise.
8. For an entertaining and nontechnical account of $1/f$ noise, see Gardner (1978), Martin Gardner's April 1978 *Mathematical Games* column in *Scientific American*. There is also no lack of more technical literature; Milotti (2001), for example, has 84 references, including some on its relation to fractals.

9. Historians of computer music should know that the program was written by composer Godfrey Winham.

10. Or, I suppose, scanned and converted to text by machine (a rather clumsy paper-tape reader).

11. For racing fans, that's one bit per furlong.

Chapter 3 / Signal Standardization

1. For an early and characteristic example of Arthur's writings on the subject, see Lo (1961).

2. At least space and time are, as far as we can tell, continuous. The idea that the universe may be, at bottom, discrete, and even itself a computer, is an old one and beyond our scope here. However, if it is discrete, it is so at an extremely small scale, or the fact would have been discovered by now in physical experiments. In the present discussion, we can assume that the world is, in fact, analog. For some informal discussion of the idea, see Feynman (1982), a paper we have more to say about later.

3. Sometimes also called signal *restoration*.

4. My approach to computer logic in this chapter follows Schaffer (1988). In that textbook, Schaffer does show in some detail how modern computers can be constructed step-by-step, hierarchically, from the one building block of the valve. This is not the usual text for an introductory course, but it is a tour de force of demystification.

5. Thomson (1897).

6. Electrons are negatively charged, and the direction of current flow is defined in general by the direction that positive charge flows, so the *current* flow in a vacuum-tube diode is actually from the plate to the filament.

7. De Forest (1908).

8. In the jargon of the computer science trade, we are putting the valve inside a "black box."

9. This is common sense, but it's also an example of De Morgan's law. When you apply NOT to a logical expression, it NOTs the variables, turns ANDs into ORs, and ORs into ANDs.

10. We follow again the parsimonious plan of Schaffer (1988), but rather more crudely.

11. Zuse (1993).

12. Lavington (1980), pp. 6–7.

13. See Goldstine (1972), for example, for a first-hand account.

14. Zuse (1993), pp. 62–63.

Chapter 4 / Consequential Physics

1. As we discussed in connection with $1/f$ noise in chapter 2.

2. Quoted in Hermann (1971), p. 23. The letter was written to Robert Williams Wood in 1931.

3. Hermann (1971), p. 11.

4. There are situations where energy *can* take on a continuous range of values, but there is no reason to worry about that here.

5. Translation from Arons and Peppard (1965).
6. In the appropriate units, a bookkeeping detail we'll gloss over here.
7. Actually, it is the product of position and *momentum* that is bounded from below by the uncertainty principle, and the principle is, in fact, much more general. However, for our purposes, momentum is mass times velocity, and we can talk about momentum and velocity interchangeably.
8. From Gillespie (1970), exercise 61, p. 108. Gillespie's book is an introductory textbook that uses undergraduate calculus, but if you have the background it is an excellent introduction to the essential structure of quantum mechanics and a model of clarity.
9. Lightning bolts are, of course, a different matter.
10. The discussion here is highly, but I hope not criminally, simplified. It takes half a semester's worth of quantum mechanics to explain the rules governing the arrangement of electrons in atoms.
11. And two neutrons, with no charge. We won't worry about neutrons here.
12. If you remember freshman chemistry, these are called *covalent* bonds.
13. Recall that like charges repel and opposite charges attract.
14. The transistor can thus be either type n-p-n or p-n-p.

Chapter 5 / Your Computer Is a Photograph

1. Feynman (1960).
2. A *microphotograph* is a photograph that is greatly reduced in size. A *photomicrograph* is a photograph of a very small object.
3. Stevens (1968).
4. Technion, Israel Institute of Technology (2015).
5. They actually report one complete Hebrew bible in a 0.5×0.5 mm square, about the size of a grain of sugar.
6. Stevens (1968).
7. Günther (1962).
8. A μm, or *micron*, is 10^{-6} m, a millionth of a meter. A nm, or nanometer, is 10^{-9} m, a billionth of a meter.
9. Moore (1965).
10. Gamow (1947).
11. In Asimov (1971).
12. At least they did on the war bonds my grandmother gave me.

Chapter 6 / Music from Bits

1. *IBM 704 Manual of Operation* (1954).
2. If you've wondered about the term "core dump," it stems from this era. I expect future generations may be puzzled by such similar anachronisms as "dial tone," "carbon copy" (as in "cc:"), or even "film."
3. I told this story in an article about the history of digital signal processing; see Steiglitz (2005). The piece has more technical detail and a more general perspective that I return to later.
4. We make acquaintance with another isomorphism at the end of this chapter.
5. See, for example, Yu (1984).

6. A signal with *absolutely* zero frequency content beyond a certain point is a mathematical abstraction. In practice we ignore frequency content when its level is so small that it is swamped by unavoidable noise.
7. Feynman (2006).
8. See Steiglitz (1996), where I present more or less the same argument. The book is meant as a gentle introduction to DSP, but does require plenty of mathematical background.
9. I'm not being familiar; Nyquist was born in Sweden, and "Harry" was his given name.
10. At least in theory and with a perfectly accurate sampling process.
11. Shannon also points out that the result was already known to mathematicians, and references Whittaker (1935).
12. See Steiglitz (1965).

Chapter 7 / Communication in a Noisy World

1. Khinchin (1957), p. 30.
2. For example, Cover and Thomas (1991).
3. My argument is after Raisbeck (1964), a classic little book, full of insights and intuitively appealing examples.
4. A trillion is 10^{12}.
5. If you want to check my arithmetic on your smartphone (made of microscopic transistors, I remind you), the information content is the log (base 2) of a trillion, minus the log of one less than a trillion.
6. To conform to the proper mathematical usage, we should be using the term *expected value* instead of *average*, but we will ignore the distinction.
7. Boltzmann had a hard time of it, ultimately hanging himself in despair over the resistance met by his theory. It is perhaps some small compensation that his tombstone in Vienna bears the engraved equation "$S = k \log W$," which is a way of writing the entropy of a gas, say, with W equally likely states. This is analogous to the entropy of a horse race with W horses all equally likely to win, though with an astronomical number of horses.
8. Gamow (1947).
9. For a recent specimen of worthwhile and entertaining writing about things entropic, see Ben-Naim (2015).
10. The word *code*, in this sense, means a translation of information from one form to another. The word is also used in another sense to refer to the instructions that computers execute.
11. Sometimes also *Shannon's second theorem*. The first is a noiseless coding theorem.
12. There have been, in fact, a succession of proofs, of increasing rigor, strength, and elegance.
13. From the standard textbook of its era, Gallager (1968), p. 12.

Chapter 8 / Analog Computers

1. This is easy to see by noting that when the 19-tooth gear turns 235 times, a total of 19 × 235 of its teeth will engage with those of the larger gear. The total number of teeth engaged by each gear must be the same, or the teeth

wouldn't mesh. Therefore, the 235-tooth gear must turn 19 times in the same period.

2. See Price (1974) for an early and detailed account of the find, initial reconstructions and interpretations, and extensive historical background on gearing and clockwork.

3. Recall Arthur C. Clarke's well-known third law: "Any sufficiently advanced technology is indistinguishable from magic."

4. Because both sides of the pulley must trace out the same distance.

5. Feynman (1988), p. 94.

6. Note that counting on one's fingers should, by our definition, be regarded as digital computation.

7. Phillips became well known later in his career for the *Phillips curve*, relating unemployment and inflation.

8. I follow the description in Swade (1995). The author, Doron Swade, was, when he wrote this piece, senior curator of computing and information technology at the Science Museum in London, where a Financephalograph is on permanent display.

9. Thomson (Lord Kelvin) (1878).

10. For Wilbur's paper, see Wilbur (1936). The quotation is from the MIT Museum (2011).

11. Before the term *computer* came to mean a machine that computes, it meant a human who computes. By the 1940s, important scientific calculations, such as those needed for the Manhattan Project, were assigned to roomfuls of computers seated before their mechanical calculators.

12. Püttmann (2014).

13. Thomson (Lord Kelvin) (1878), p. 483n.

14. Wilbur used more than one thousand ball-bearing pulleys, all, I imagine, well oiled. His acknowledgment of support from the president of the Singer Sewing Machine Company hints, perhaps, at dreams of a technology that was never to be.

15. For an early, extensive survey of machines for solving all kinds of equations, not just simultaneous linear ones, see Frame (1945). For machines for solving problems that involve simultaneous equations where we impose constraints on the variables (*linear programming problems*), see Sinden (1959).

16. For a history of the problem, a discussion of its name, and a wealth of material on many versions of it, see Hwang et al. (1992).

17. Courant and Robbins (1996), originally published in 1941 and now in a second edition with a new chapter by Ian Stewart on recent developments. Playing with soap bubbles and films is an enjoyable (but messy) science project for children, and, for that matter, adults as well.

18. See Aaronson (2005) for a report on experiments along these lines, as well as a wide-ranging discussion of the more general question of the relative power of analog and digital computing.

19. Also called *slopes* or, more mathematically speaking, *derivatives*.

20. See Bromley (1990) for an excellent survey of early mechanical computing devices, and analog computers in general.

21. If you are not familiar with the notation $f(x)$, just think of $f(x)$ as "something that depends on x."

22. Shannon (1941).

Chapter 9 / Turing's Machine

1. For some history of the embroidery industry in the New Jersey environs of Manhattan, and a chronicle of its decline, see Pristin (1998).
2. These estimates are from Essinger (2004), who has more on Jacquard, his predecessors, and the influence of his punched-card control on the development of computers.
3. Essinger (2004).
4. See Babbage (1994), the closest we have to an autobiography.
5. Quoted in the introduction to Babbage (1994), which cites Babbage (1989a).
6. Using what is called the *calculus of finite differences*. Miller (1960) is a crisp text at an undergraduate level.
7. See, for example, Hyman (1982) and Essinger (2004).
8. See http://www.computerhistory.org/babbage/engines. (Accessed May 15, 2017).
9. See Collier (1970), a PhD thesis with very valuable technical and chronological details drawn from Babbage's manuscripts and "Scribbling Books," and a source I rely on in this account.
10. For the patent, see Rosenberger (1960).
11. Menabrea (1842).
12. But see the fine book-length treatments of Lovelace now available, for example, Essinger (2014) and Woolley (2015). Much of her extensive correspondence with Babbage survives, some of it quite technical, some fanciful. See especially her long and revealing letter of August 14, 1843—after knowing Babbage only a bit more than two months; it is reprinted in full as Appendix 2 of Essinger (2004). She suffered and died young; she was, in a word, Byronic.
13. Lovelace's translation of Menabrea's paper, with her notes, was published as Menabrea (1843) and is reprinted in full in Babbage (1989b).
14. It is not a surprise that Bernoulli numbers play a central role in the calculus of finite differences, which we met in connection with Babbage's difference engines. See, for example, Miller (1960).
15. Some definitions of a Turing machine use a tape that extends indefinitely to the right but has a definite, fixed end at the left. There is also the question of how the tape is initially prepared, but we always assume that it is primed with the input data for any particular problem and that there is only a finite amount of that input data. In fact, many of the details in our construction differ from computer scientist to computer scientist. As we discuss in the next chapter, it turns out that none of these details matters.
16. Turing (1936) uses the term *configuration* for *state*.
17. Actually, the contents of the cells of a cellular automaton are usually all updated simultaneously. But these details don't concern us here. For much more about cellular automata, see Wolfram (2002). He describes there the remarkable "Rule 110" machine, which is, in theory, as powerful as a Turing machine but scans only three cells with its head.
18. von Neumann (1966).

Chapter 10 / Intrinsic Difficulty

1. The study of computational complexity in any depth is usually introduced in a one-semester course at the advanced undergraduate level. My informal survey in this chapter is along the lines of the moderately technical Papadimitriou and Steiglitz (1982). For a standard undergraduate text that is devoted only to computational complexity, see Papadimitriou (1994), Sipser (1997), or, at the graduate level, Arora and Barak (2009).

2. The Soviet-born, now US émigré Leonard Levin independently developed Cook's main result, but published it later. What we call "Cook's theorem" is often also called the "Cook-Levin theorem."

3. Technically, the *asymptotic* time complexity as the problem instance size grows large.

4. This definition of *simulate* suits our purposes in this chapter but is appropriate only for digital computers, when the question of precision doesn't come up. When we use the term in connection with analog machines, as we do in the next chapter, we mean that the simulating machine matches the behavior of the simulated machine within a given precision without an exponential blowup in computing time.

5. This language for logic was invented by George Boole in the mid-nineteenth century, and the system for manipulating its expressions is called *Boolean algebra*. Claude Shannon made good use of it to design circuits for computer logic and telephone switching; see, for example, Shannon (1938), which is abstracted from his master's thesis.

6. As usual, polynomial in the length of a reasonable size parameter; in a SAT problem, the length of the input CNF formula.

7. The name NP for this class of problems comes from *nondeterministic polynomial* and refers to the fact that an imaginary kind of Turing machine, called a *nondeterministic* Turing machine, can essentially guess the certificate.

8. The website http://www.math.uwaterloo.ca/tsp/ (accessed September 26, 2017) is a rich source of information about the TSP and its history, including details of the Procter and Gamble contest. See also Cook (2012) for an entertaining tour and Applegate et al. (2011) for a comprehensive treatment of current computer methods for solution.

9. I give here, of course, only a rough outline of the main argument; for details, see, for example, Papadimitriou and Steiglitz (1982).

10. Actually, the argument summarized here uses CNFs that can have more than three variables in a factor, but the more general problem is polynomially reducible to 3-SAT.

11. Karp (1972).

12. There is a technical difference between the kind of reduction used by Karp and that used by Cook, but the difference need not concern us here.

Chapter 11 / Searching for Magic

1. See Garey and Johnson (1979), an early but excellent and still widely used collection of NP-complete problems. PARTITION is also one of the original 21 problems in Karp (1972).

2. I rigged this example so the sum of the first four integers is equal to the sum of the last three. Life is not always this easy. Imagine, for example, a set with a few thousand integers that cover a much larger range.

3. I described this machine in Steiglitz (1988), where I debunk it as I do here.

4. A fact that was appreciated early on, and noted, for example, in von Neumann (1958). For those readers who know some calculus, the trick is to use integration by parts. Radio receivers don't use multipliers or integrators, but use a third way to implement mixing, passing the signals through a nonlinear circuit device like a diode.

5. Vergis et al. (1986). The spiritual ancestry of Vergis's machine is also noted by the esteemed Dervish Hazaroglu (quoted in Papadimitriou (2005), p. 262).

6. Lee (1999).

7. Main (1994, 2007).

8. Aaronson (2005).

9. Markov (2015) provides a critique of this paper. The basic idea of the machine in Traversa et al. (2015) is actually the same as that used in the PARTITION machine described above.

10. Any number with a decimal representation that terminates can be written down by a Turing machine, so we are necessarily talking about real numbers with a nonterminating decimal expansion.

11. The Church-Turing thesis is often called Church's thesis. It depends on who is doing the calling.

12. See, for example, the collection of papers in Olszewski et al. (2006).

13. Also called *Strong Church's Thesis* in Vergis et al. (1986) and the *Strong Form of the Church-Turing Thesis* in Arora and Barak (2009).

14. In the world of theoretical computer science, the class of problems naturally corresponding to P when we allow randomness is called BPP, for *bounded-error probabilistic polynomial time*.

15. The physics behind this hypothetical experiment has been verified convincingly in the laboratory many times and under a variety of conditions.

16. Peebles (1992), pp. 252ff., attributes a very similar version of Bell's hypothetical experiment to Eugene Wigner (with no citation). For a history of the slow recognition of Bell's theorem, including Feynman's early qualms about the worldview of quantum mechanics, see Freire (2006).

17. Einstein et al. (1935).

18. In Nabokov (1955), endnote, he comments on the illusive nature of reality with a parenthetical phrase: "'reality' (one of the few words which mean nothing without quotes)." It seems especially important to respect his warning when discussing quantum mechanics.

19. "Black box" is the traditional term for such a device among engineers and scientists.

20. Named *bra-ket* notation by the great physicist Paul Dirac.

21. The technical details are explained in many places, one of the clearest being the class notes of Preskill (1998).

22. Shor (1994).

23. Rivest et al. (1978).

24. The more advanced reader is referred to the well-known oracle results in Baker et al. (1975) and Bennett and Gill (1981).

25. See, for example, Bennett et al. (1997).
26. For a review of general computing systems that can be built at the biomolecular level, see Benenson (2012).
27. An enormous amount is known about neurons and how they interact. I am indebted to G. Y. Buzsáki for confirming the idea that there are neuron types that can, under the right circumstances, act as simple gates. For a peek at the highly specialized literature on the subject, see, for example, Freund and Buzsáki (1996).
28. I am ignoring here the possibility that the brain might operate along quantum-mechanical lines, a speculative, controversial, and, in my view, unlikely prospect. We return to this question in the next chapter.
29. I wrote about something like the Singularity as a brash undergraduate in Steiglitz (1959), and it was no doubt not a new idea then.

Chapter 12 / The Internet, Then the Robots

1. I rely in the following on Hutt et al. (1993) for an excellent history of Bell's photophone, which we discuss next.
2. Bell was evidently thrilled by the whole idea and considered this his greatest invention. Hutt et al. include a report that he considered naming his newly born second daughter "Photophone," but reconsidered, perhaps mercifully for the girl.
3. As of this writing, three times a week, at http://technews.acm.org/. The ACM is the dominant computer professional organization. The use of the word *machinery* in its name may seem quaint to some, but I especially like it because it admits a broad view of what can be considered a computer.
4. I use the very common acronym AI.
5. I often use the word *brain* in this chapter to refer to the *human* brain. In all modesty, it is the best one we know of.
6. We use the term *neuron* to refer to both the natural (biological) and the synthetic (software) versions.
7. A job that is now under challenge by machines.
8. Machines seem a long way from dealing with this problem.
9. The classic and influential paper McCulloch and Pitts (1943) is representative of the thinking mid-twentieth century.
10. See Skinner (1960), where he gives his candid, and in some ways prophetic, account of the project. It seems that I can't avoid the Second World War: it was a time of great tragedies, but also a time of great scientific stimulation, and in many ways the beginning of a new era.
11. The idea never reached fruition. Incidentally, the pigeons were trained with "a target in New Jersey consisting of a stirrup shaped pattern bulldozed out of the sandy soil near the coast." He laments that in the end, "we had to show, for all our trouble, only a loftful of curiously useless equipment and a few dozen pigeons with a strange interest in a feature of the New Jersey coast."
12. Speaking of rewards, Skinner notes that "pigeons were said to find hemp seed particularly delectable."
13. Rothganger (2017).
14. Gomes (2017).

15. It may take many operations to adjust each weight, so this is an optimistic estimate.
16. See the paper by Avarguès-Weber and Giurfa (2013), which is short and sweet.
17. Penrose (1989).
18. For a clear and pointed critique, see Koch and Hepp (2006).
19. Čapek (1923). For another Čapek work that depicts interspecies conflict, but with newts instead of robots, see Čapek (1999). A delightful and eerie masterpiece.
20. Dick (1996). The 1982 film *Blade Runner*, directed by Ridley Scott, is loosely based on the novel and uses the term *replicant* instead.
21. As put by Weisberg (2014), a compact survey of current academic work on consciousness.
22. Chalmers (1995).
23. Chalmers (1996).
24. So termed by Flanagan (1991), after the 1960s rock band Question Mark and the Mysterians.
25. Monod (1971), p. 146.
26. In Searle (1980), a paper famous for introducing the "Chinese room" argument, which we discuss below.
27. Churchland's well-aimed academic barb at Penrose and Hameroff; see Churchland (1998).
28. Chalmers (1996).
29. Searle (1980).
30. Here I follow Chalmers (1996).
31. See, for example, Ridley (1996), who finds the origins of our instincts for mutual aid and cooperation in our evolutionary history.
32. Asimov (1950).

Bibliography

Aaronson, S. 2005. NP-complete problems and physical reality. *SIGACT News Complexity Theory column*, March. Available at http://www.scottaaronson .com/papers/npcomplete.pdf, accessed August 26, 2016.

Applegate, D. L., R. E. Bixby, V. Chvatal, and W. J. Cook. 2011. *The Traveling Salesman Problem: A Computational Study*. Princeton University Press, Princeton, NJ.

Arons, A. B., and M. B. Peppard. 1965. Einstein's proposal of the photon concept—A translation of the *Annal der Physik* paper of 1905. *Amer. J. Phys.*, 33(5):367–374, May.

Arora, S., and B. Barak. 2009. *Computational Complexity: A Modern Approach*. Cambridge University Press, Cambridge, UK.

Asimov, I. 1950. Runaround. In *I, Robot*. Fawcett Crest, Greenwich, CT. Originally published 1942.

Asimov, I. 1971. *The Stars in Their Courses*. Doubleday & Company, New York. Originally published in *Magazine of Fantasy and Science Fiction*, May 1969.

Avarguès-Weber, A., and M. Giurfa. 2013. Conceptual learning by miniature brains. *Proc. Royal Soc. B*, 280, December 7, issue 1772, doi: 10.1098/rspb.2013.1907.

Babbage, C. 1989a. The science of number reduced to mechanism. In M. Campbell-Kelly, editor, *The Works of Charles Babbage*. W. Pickering, London, vol. 2, pages 15–32. Originally published 1822.

Babbage, C. 1989b. *Science and Reform: Selected Works of Charles Babbage*. Cambridge University Press, Cambridge, UK. Introduction and discussion by A. Hyman.

Babbage, C. 1994. *Passages from the Life of a Philosopher*. Rutgers University Press, New Brunswick, NJ, and IEEE Press, New York. Originally published 1864.

Baker, T., J. Gill, and R. Solovay. 1975. Relativizations of the P = ?NP question. *SIAM Journal on Computing*, 4(4):431–442.

Bar-Lev, A. 1993. *Semiconductors and Electronic Devices*. Prentice-Hall, Englewood Cliffs, NJ, third edition.

Basdevant, J-L. 2007. *Lectures in Quantum Mechanics*. Springer, New York.

Bell, J. S. 1964. On the Einstein Podolsky Rosen paradox. *Physics*, 1(3):195–200.

Ben-Naim, A. 2015. *Information, Entropy, Life and the Universe: What We Know and What We Do Not Know*. World Scientific, Hackensack, NJ.

Benenson, Y. 2012. Biomolecular computing systems: Principles, progress and potential. *Nature Reviews Genetics*, 13(7):455–468.

Bennett, C. H., and J. Gill. 1981. Relative to a random oracle A, $P^A \neq NP^A \neq$ co-NP^A with probability 1. *SIAM Journal on Computing*, 10(1):96–113.

Bennett, C. H., E. Bernstein, G. Brassard, and U. Vazirani. 1997. Strengths and weaknesses of quantum computing. *SIAM Journal on Computing*, 26(5): 1510–1523.

Blahut, R. E. 1987. *Principles and Practice of Information Theory*. Addison-Wesley, Reading, MA.

Blausen.com staff. 2014. Medical gallery of Blausen Medical. *WikiJournal of Medicine*, 1(2), doi: 10.15347/wjm/2014.010. Available at https://upload .wikimedia.org/wikiversity/en/7/72/Blausen_gallery_2014.pdf, accessed September 27, 2017.

Bromley, A. G. 1990. Analog computing devices. In W. Aspray, editor, *Computing before Computers*. Iowa State University Press, Ames. Available at http://ed-thelen.org/comp-hist/CBC-Ch-05.pdf, accessed September 13, 2016.

Bush, V. 1931 The differential analyzer: A new machine for solving differential equations. *J. Franklin Inst.*, 212:447–488.

Čapek, K. 1923. *R.U.R. (Rossum's Universal Robots)*. Doubleday, Page & Co., Garden City, NY. Translated from the Czech by P. Selver.

Čapek, K. 1999. *War with the Newts*. Catbird Press, North Haven, CT. Translated from the Czech by E. Osers; originally published 1936.

Chalmers, D. J. 1995. Facing up to the problem of consciousness. *Journal of Consciousness Studies*, 2(3):200–219.

Chalmers, D. J. 1996. *The Conscious Mind: In Search of a Fundamental Theory*. Oxford University Press, Oxford, UK.

Churchland, P. S. 1998. Brainshy: Nonneural theories of conscious experience. In S. R. Hameroff, A. W. Kaszniak, and A. C. Scott, editors, *Towards a Science of Consciousness II: The Second Tucson Discussions and Debates*, pages 109–126. MIT Press, Cambridge, MA.

Collier, B. 1970. *Little engines that could've: The calculating machines of Charles Babbage*. PhD thesis, Harvard University, Cambridge, MA, August.

Cook, S. A. 1971. The complexity of theorem proving procedures. In *Proc. 3rd ACM Symp. on the Theory of Computing*, pages 151–158.

Cook, W. J. 2012. *In Pursuit of the Traveling Salesman: Mathematics at the Limits of Computation*. Princeton University Press, Princeton, NJ.

Courant, R., and H. Robbins. 1996. *What Is Mathematics?—An Elementary Approach to Ideas and Methods*. Oxford University Press, New York, second edition. Revised by Ian Stewart.

Cover, T. M., and J. A. Thomas. 1991. *Elements of Information Theory*. John Wiley, New York.

De Forest, L. 1908. Space telegraphy. US Patent no. 879,532; filed January 29, 1907; issued February 18, 1908, Available at. http://history-computer.com/Library/US879532.pdf, accessed April 19, 2016.

Deutsch, D., and R. Jozsa. 1992. Rapid solution of problems by quantum computation. *Proceedings of the Royal Society of London A: Mathematical, Physical and Engineering Sciences*, 439(1907):553–558.

Dick, P. K. 1996. *Do Androids Dream of Electric Sheep?* Ballantine Books, New York. Originally published 1968.

Einstein, A., B. Podolsky, and N. Rosen. 1935. Can quantum-mechanical description of physical reality be considered complete? *Physical Review*, 47(10):777.

Essinger, J. 2004. *Jacquard's Web: How a Hand-Loom Led to the Birth of the Information Age*. Oxford University Press, Oxford, UK.

Essinger, J. 2014. *Ada's Algorithm: How Lord Byron's Daughter Ada Lovelace Launched the Digital Age*. Melville House, Brooklyn, NY.

Feynman, R. P. 1960. There's plenty of room at the bottom. *Caltech Engineering and Science*, 23(5):22–36, February. Transcript of a talk given December 29, 1959, at the annual meeting of the American Physical Society at the California Institute of Technology. Available at http://www.zyvex.com/nanotech/feynman.html, accessed June 19, 2016.

Feynman, R. P. 1982. Simulating physics with computers. *Int. J. Theor. Physics*, 21(6/7):467–488. Keynote address delivered at the First MIT Physics of Computation Conference, May 6–8 1981.

Feynman, R. P. 1988. *What Do YOU Care What Other People Think?: Further Adventures of a Curious Character, as Told to Ralph Leighton*. W. W. Norton & Company, New York.

Feynman, R. P. 2006. *QED: The Strange Theory of Light and Matter*. Princeton University Press, Princeton, NJ. Originally published 1985.

Flanagan, O. J. 1991. *The Science of the Mind*. MIT Press, Cambridge, MA, second edition.

Frame, J. S. 1945. Machines for solving algebraic equations. *Math. Comp.*, 1: 337–353.

Freeth, T., et al. 2006. Decoding the ancient Greek astronomical calculator known as the Antikythera mechanism. *Nature*, 444(30):587–591, November.

Freire, O. 2006. Philosophy enters the optics laboratory: Bell's theorem and its first experimental tests (1965–1982). *Studies in History and Philosophy of Science Part B: Studies in History and Philosophy of Modern Physics*, 37(4):577–616. Available at https://arxiv.org/ftp/physics/papers/0508/0508180.pdf.

Freund, T. F., and G. Y. Buzsáki. 1996. Interneurons of the hippocampus. *Hippocampus*, 6(4):347–470.

Friedrichs, H. P. 2003. *Instruments of Amplification: Fun with Homemade Tubes, Transistors, and More*. Self-published; ISBN 0-9671905-1-7. Available at http://www.hpfriedrichs.com/mybooks/mybooks.htm, accessed October 9, 2017.

Gallager, R. G. 1968. *Information Theory and Reliable Communication*. John Wiley, New York.

Gamow, G. 1947. *One, Two, Three … Infinity: Facts & Speculations of Science*. Viking Press, New York. Revised 1961; reprinted Dover, 1988.

Gardner, M. 1978. White and brown music, fractal curves and one-over-f fluctuations (Mathematical Games column). *Sci. Amer.*, 238:16–32, April.

Garey, M. R., and D. S. Johnson. 1979. *Computers and Intractability*. Freeman, San Francisco.

Gillespie, D. T. 1970. *A Quantum Mechanics Primer*. International Textbook Co., Scranton, PA.

Goldstine, H. H. 1972. *The Computer: From Pascal to von Neumann*. Princeton University Press, Princeton, NJ.

Gomes, L. 2017. The neuromorphic chip's make-or-break moment. *IEEE Spectrum*, 54(6):53–57, June.

Günther, A. 1962. Microphotography in the library. *Unesco Bulletin for Libraries*, XVI(1), January–February, Item I.

Hameroff, S., and R. Penrose. 1996. Orchestrated reduction of quantum coherence in brain microtubules: A model for consciousness. *Mathematics and Computers in Simulation*, 40(3–4):453–480.

Hameroff, S. R. 1994. Quantum coherence in microtubules: A neural basis for emergent consciousness? *Journal of Consciousness Studies*, 1(1):91–118.

Hecht, J. 2016. Great leaps of light. *IEEE Spectrum*, 53(2):28–53, February.

Hermann, A. 1971. *The Genesis of Quantum Theory (1899–1913)*. MIT Press, Cambridge, MA. Translated by C. W. Nash.

Horowitz, P., and W. Hill. 1980. *The Art of Electronics*. Cambridge University Press, Cambridge, UK.

Hutt, D. L., K. J. Snell, and P. A. Bélanger. 1993. Alexander Graham Bell's photophone. *Optics & Photonics News*, 4(6):20–25, June.

Hwang, F. K., D. S. Richards, and P. Winter. 1992. *The Steiner Tree Problem*. North-Holland, Amsterdam.

Hyman, A. 1982. *Charles Babbage: Pioneer of the Computer*. Princeton University Press, Princeton, NJ.

IBM 704 Manual of Operation. 1954. IBM, New York.

Irwin, W. 2013/2014. The Cambridge Meccano differential analyser no. 2. *Computer Resurrection: Bulletin of the Computer Conservation Society* (64), Winter. Available at http://www.computerconservationsociety.org/resurrection/res64 .htm, accessed September 15, 2017.

Isenberg, C. 1976. The soap film: An analogue computer. *Amer. Sci.*, 64(5):514–518, September–October.

Johnson, J. B. 1928. Thermal agitation of electricity in conductors. *Phys. Rev.*, 32:97–109, July.

Karp, R. M. 1972. Reducibility among combinatorial problems. In R. E. Miller and J. M. Thatcher, editors, *Complexity of Computer Computations*, pages 85–103. Springer, New York.

Khan, Ustad Imrat. 1991. Ajmer. Water Lily Acoustics compact disc WLA-ES-17-CD (Surbahār and Sitār, Shafaatullah Khan, Tablā).

Khinchin, A. I. 1957. *Mathematical Foundations of Information Theory*. Dover, New York. Translated from the Russian by R. A. Silverman and M. D. Friedman.

Koch, C., and K. Hepp. 2006. Quantum mechanics in the brain. *Nature*, 440(7084):611–612, March 30.

Lavington, S. H. 1980. *Early British Computers: The Story of Vintage Computers and the People Who Built Them*. Manchester University Press, Manchester, UK.

Lee, F. 1999. Physical manifestation of NP-completeness in analog computer devices. Master's thesis, MIT, Cambridge, MA.

Lo, A. 1961. Some thoughts on digital components and circuit techniques. *IRE Transactions on Electronic Computers*, EC-10(3):416–425, September.

Main, M. G. 1994. Analog solution of NP-complete problems. Technical report CU-CS-700-94, Computer Science Department, University of Colorado, Boulder, paper 668. Available at http://scholar.colorado.edu/csciw_techreports /668.

Main, M. G. 2007. Building a prototype analog computer for exact-1-in-3-SAT. Technical Report CU-CS-1035-07, Computer Science Department, University of Colorado, Boulder, paper 967. Available at http://scholar .colorado.edu/csci_techreports/967, accessed October 2, 2016.

Markland, E., and R. F. Boucher. 1971. Fundamentals of fluidics. In A. Conway, editor, *A Guide to Fluidics*. Macdonald & Co. Ltd., London.

Markov, I. L. 2015. A review of "Mem-computing NP-complete problems in polynomial time using polynomial resources." Review of Traversa et al. (2015). Available at e-print archive https://arxiv.org/abs/1412.0650, accessed January 22, 2018.

McCulloch, W. S., and W. S. Pitts. 1943. A logical calculus of the ideas immanent in nervous activity. *Bulletin of Mathematical Biophysics*, 5(4):115–133.

Menabrea, L. F. 1842. Notions sur la machine analytique de M. Charles Babbage. *Bibliothèque Universelle de Genève*, 41:352–376.

Menabrea, L. F. 1843. Sketch of the analytical engine invented by Charles Babbage from the Bibliothèque Universelle de Genève, October, 1842, no. 82. *Scientific Memoirs*, iii:666–731. Translated with notes by Ada Augusta Lovelace.

Miller, K. S. 1960. *An Introduction to the Calculus of Finite Differences and Difference Equations*. Holt, New York.

Milotti, E. 2001. 1/f noise: A pedagogical review. Availaable at http://arxiv .org/abs/physics/0204033v1, Invited talk to E-GLEA-2, Buenos Aires, September 10–14. accessed April 19, 2016.

MIT Civil and Environmental Engineering Newsletter. Original still missing, but copy of Wilbur mechanical calculator reappears in Tokyo. Winter 2001.

MIT Museum. 2011. Wilbur machine. Photo from a nomination for the MIT 150 Exhibition, which opened January 8, 2011. Available at http://museum .mit.edu/nom150/entries/1422, accessed August 29, 2016.

Monod, J. 1971. *Chance and Necessity: An Essay on the Natural Philosophy of Modern Biology*. Knopf, New York. Translated from the French by A. Wainhouse.

Moore, G. E. 1965. Cramming more components onto integrated circuits. *Electronics Magazine*, 38(8), April 19. Reprinted in *IEEE Solid-State Circuits Soc. Newsletter*, September 2006, 33–35.

Nabokov, V. 1955. *Lolita*. G. P. Putnam, New York.

Newburgh, R., J. Peidle, and W. Rueckner. 2006. Einstein, Perrin, and the reality of atoms: 1905 revisited. *Am. J. Phys.*, 74(6):478–481, June.

Nyquist, H. 1928a. Certain topics in telegraph transmission theory. *Trans. AIEE*, 47:617–644, April.

Nyquist, H. 1928b. Thermal agitation of electrical charge in conductors. *Phys. Rev.*, 32:110–113, July.

Olszewski, A., J. Woleński, and R. Janusz, editors. 2006. *Church's Thesis after 70 Years*. Ontos Verlag, Heusenstamm, Germany.

Oltean, M. 2008. Solving the Hamiltonian path problem with a light-based computer. *Natural Computing*, 7(1):57–70.

Pakkenberg, B., D. Pelvig, L. Marner, M. J. Bundgaard, H. J. G. Jørgen, J. R. Nyengaard, and L. Regeur. 2003. Aging and the human neocortex. *Experimental Gerontology*, 38(1):95–99.

Papadimitriou, C. H. 1994. *Computational Complexity*. Addison-Wesley, Reading, MA.

Papadimitriou, C. H. 2005. *Turing (A Novel about Computation)*. MIT Press, Cambridge, MA.

Papadimitriou, C. H., and K. Steiglitz. 1982. *Combinatorial Optimization: Algorithms and Complexity*. Prentice-Hall, Englewood Cliffs, NJ. Reprinted with corrections, Dover, New York, 1996.

Peebles, P. J. E. 1992. *Quantum Mechanics*. Princeton University Press, Princeton, NJ.

Penrose, R. 1989. *The Emperor's New Mind: Concerning Computers, Minds, and the Laws of Physics*. Oxford University Press, New York.

Preskill, J. 1998. Lecture notes for Physics 229: Quantum Information and Computation. California Institute of Technology, September. Available at https://www.lorentz.leidenuniv.nl/quantumcomputers/literature/preskill_1_to_6.pdf, accessed October 15, 2016.

Press, W. H. 1978. Flicker noises in astronomy and elsewhere. *Comments Astrophys.*, 7(4):103–119.

Price, D. J. 1974. Gears from the Greeks: The Antikythera mechanism, a calendar computer from ca. 80 B.C. *Trans. Amer. Phil. Soc.*, 64(7):1–70.

Pristin, T. 1998. In New Jersey, a delicate industry unravels. *New York Times*, January 3. Available at http://www.nytimes.com/1998/01/03/nyregion/in-new-jersey-a-delicate-industry-unravels.html, accessed May 18, 2017.

Püttmann, T. 2014. Kelvin: A simultaneous calculator. Available at http://www.math-meets-machines.de/kelvin/simcalc.pdf, accessed August 29, 2016.

Raisbeck, G. 1964. *Information Theory: An Introduction for Scientists and Engineers*. MIT Press, Cambridge, MA.

Ridley, M. 1996. *The Origins of Virtue*. Penguin Books, New York.

Rivest, R. L., A. Shamir, and L. Adleman. 1978. A method for obtaining digital signatures and public-key cryptosystems. *Communications of the ACM*, 21(2): 120–126.

Rosenberger, G. B. 1960. Simultaneous carry adder. US Patent 2,966,305; filed August 16, 1957; published December 27, 1960. Available at http://www.google.com/patents/US2966305, accessed May 18, 2017.

Rothganger, F. 2017. The dawn of the real thinking machine. *IEEE Spectrum*, 54(6):22–25, June.

Roy, S., and A. Asenov. 2005. Where do the dopants go? *Science*, 309, July 15.

Russell, B. 2009. *ABC of Relativity*. Routledge, New York. Originally published 1941.

Schaffer, C. 1988. *Principles of Computer Science*. Prentice-Hall, Englewood Cliffs, NJ.

Searle, J. R. 1980. Minds, brains, and programs. *Behavioral and Brain Sciences*, 3(3): 417–424.

Shannon, C. E. 1938. A symbolic analysis of relay and switching circuits. *Trans. Amer. Inst. of Elect. Eng.*, 57(12):713–723.

Shannon, C. E. 1941. Mathematical theory of the differential analyzer. *J. Math. & Phys.*, 20:337–354, April.

Shannon, C. E. 1948. A mathematical theory of communication. *Bell Sys. Tech. J.*, 27:379–423, 623–656.

Shannon, C. E. 1949. Communication in the presence of noise. *Proc. IRE*, 37(1): 10–21, January.

Shor, P. W. 1994. Algorithms for quantum computation: Discrete logarithms and factoring. In *Proceedings of the 35th Annual Symposium on the Foundations of Computer Science*, pages 124–134.

Sinden, F. W. 1959. Mechanisms for linear programs. *Operations Research*, 7(6): 728–739, November–December.

Sipser, M. 1997. *Introduction to the Theory of Computation*. PWS, Boston.

Skinner, B. F. 1960. Pigeons in a pelican. *American Psychologist*, 15(1):28–37.

Steiglitz, K. 1959. The simulation of human activities by machine. *Quadrangle*, 29(3):23–24, January. College of Engineering, New York University. Available at http://www.cs.princeton.edu/~ken/simulation_human59.pdf.

Steiglitz, K. 1965. The equivalence of digital and analog signal processing. *Information & Control*, 8(5):455–467, October.

Steiglitz, K. 1988. Two non-standard paradigms for computation: Analog machines and cellular automata. In J. K. Skwirzynski, editor, *Performance Limits in Communication Theory and Practice*, pages 173–192. Kluwer, Dordrecht, Netherlands. NATO Advanced Study Institute on Performance Limits in Communication Theory and Practice, series E, no. 142, July 7–19, 1986.

Steiglitz, K. 1996. *A DSP Primer*. Prentice-Hall, Englewood Cliffs, NJ.

Steiglitz, K. 2005. Isomorphism as technology transfer. *Signal Processing Magazine*, 22:171–173, November.

Stevens, G. W. W. 1968. *Microphotography: Photography and Photofabrication at Extreme Resolution*. John Wiley, New York, second edition.

Swade, D. 1995. When money flowed like water. *Inc.*, September 15. Available at http://www.inc.com/magazine/19950915/2624.html, accessed August 22, 2016.

Technion, Israel Institute of Technology. 2015. And then there was Nano—The smallest bible in the world. Available at http://int.technion.ac.il/and-then-there-was-nano-the-smallest-bible-in-the-world-on-exhibit-at-the-shrine-of-the-book/, accessed May 16, 2016.

Thomson, J. J. 1897. Cathode rays. *Philosophical Magazine*, 44:293–316.

Thomson, W. (Lord Kelvin). 1878. On a machine for the solution of simultaneous linear equations. *Proc. Roy. Soc.*, xxviii:111–113, December 5.

Thomson, W. (Lord Kelvin), and P. G. Tait. 1890. *Treatise on Natural Philosophy*. Cambridge University Press, Cambridge, UK, new edition, part I.

Traversa, F. L., C. Ramella, F. Bonani, and M. Di Ventra. 2015. Memcomputing NP-complete problems in polynomial time using polynomial resources and collective states. *Science Advances*, 1(6).

Turing, A. M. 1936. On computable numbers, with an application to the Entscheidungsproblem. *Proceedings of the London Mathematical Society*, pages 230–265.

Vergis, A., K. Steiglitz, and B. D. Dickinson. 1986. The complexity of analog computation. *Mathematics and Computers in Simulation*, 28:91–113.

von Neumann, J. 1958. *The Computer and the Brain*. Yale University Press, New Haven, CT.

von Neumann, J. 1966. *Theory of Self-Reproducing Automata*. University of Illinois Press, Urbana. Edited and completed by A. W. Burks.

Weisberg, J. 2014. *Consciousness*. Polity Press, Cambridge, UK.

Whittaker, J. M. 1935. *Interpolatory Function Theory*. Cambridge University Press, Cambridge, UK, ch. IV.

Wilbur, J. B. 1936. The mechanical solution of simultaneous equations. *J. Franklin Inst.*, 222:715–724, December.

Wolfram, S. 2002. *A New Kind of Science*. Wolfram Media, Champaign, IL.

Woolley, B. 2015. *The Bride of Science: Romance, Reason and Byron's Daughter*. Pan Macmillan, London.

Yao, A. C-C. 2003. Classical physics and the Church–Turing thesis. *Journal of the ACM*, 50(1):100–105.

Yu, F. T. S. 1984. *Optics and Information Theory*. John Wiley, New York.

Zuse, K. 1993. *The Computer—My Life*. Springer-Verlag, Berlin. Originally published in German, 1984; translated by P. McKenna and J. A. Ross.

Index